Sänger

Germany's Orbital Rocket Bomber in World War II

"All men dream, but not equally. Those who dream by night in the dusty recesses of their minds wake in the day to find that it was vanity; but the dreamers of the day are dangerous men, for they may act their dream with open eyes, to make it possible."

T. E. Lawrence——*The Seven Pillars of Wisdom*

David Myhra

Schiffer Military History
Atglen, PA

Acknolwedgments

The author has benefitted in writing this book through the help of several very gifted people. They include:

- *Al Bowers* - manuscript review
- *Betsy Hertel* - wife, best friend, and supportive fan
- *Josha Hildwine* - digital images
- *Mario Merino* - digital images
- *Ian Robertson* - book editor
- *Reinhard Roeser* - archival research
- *Peter Schiffer* - publisher
- *Geoff Steele* - manuscript editor

Book Design by Ian Robertson.

Library of Congress Catalog Number: 2001095479

Printed in China.
ISBN: 0-7643-1549-8

We are interested in hearing from authors with book ideas on related topics.

Published by Schiffer Publishing Ltd.
4880 Lower Valley Road
Atglen, PA 19310
Phone: (610) 593-1777
FAX: (610) 593-2002
E-mail: Schifferbk@aol.com.
Visit our web site at: www.schifferbooks.com
Please write for a free catalog.
This book may be purchased from the publisher.
Please include $3.95 postage.
Try your bookstore first.

In Europe, Schiffer books are distributed by:
Bushwood Books
6 Marksbury Avenue
Kew Gardens
Surrey TW9 4JF
England
Phone: 44 (0) 20 8392-8585
FAX: 44 (0) 20 8392-9876
E-mail: Bushwd@aol.com.
Free postage in the UK. Europe: air mail at cost.
Try your bookstore first.

Contents

Introduction

"I remember *Vasilli Stalin*, son of the USSR's iron-fisted dictator *Josef Stalin*. He'd come frequently to *Podberese'je* (about 100 miles from Moscow). We'd all drink vodka, of course. A funny thing about young *Stalin*. No matter how much he drank...and he drank a lot...it never loosened his tongue. *Podberese'je*...was that rural area where most of us German aviation specialists ended up after we had been deported to the USSR in October, 1946. Among us were men from almost every German aircraft manufacturer. They were from *Heinkel AG*-Oranienburg and Rostock, a few from *Gothaer Waggonfabric*-Gotha, but most were from *Junkers*-Dessau. I and few others came from *Siebel*-Halle. Well, each time *Vasilli* came to visit, he'd always tell us right at the start: 'I will give to any one their weight in gold who can tell me where to find *Eugen Sänger*.' If I knew *Sänger's* whereabouts, I could have been a very rich man because at that time — in late 1946 — I weighed 170 pounds. So that would be equal to about 2,720 ounces. If gold was selling, let's say, for $200 per ounce...then I'd be worth over a half million dollars. It didn't happen because not one of us knew where *Sänger* could be found. So you see, this shows you how badly the Soviets wanted to capture alive the man whose dream was to create an orbital bomber...*Dr.-Ing. Eugen Sänger*."

Hans Motsch, Bitterfeld, East Germany,
as spoken to the author in 1988.

What was so valuable about the Austrian rocket expert *Eugen Albert Sänger* (1905-1964) that the Soviets were readily willing to give a half million or more dollars in gold just to know in what country and in what city they could find him? *Sänger* was no ordinary German aviation scientist. He was one of the most vocal and ardent advocates of developing a bi-fuel liquid rocket engine of his own design...using gasoline, oil, and liquid oxygen...capable of

A *Sänger* piloted, winged, sub-orbital reusable bi-liquid rocket bomber seen from its port side about 78 to 80 miles above Indianapolis, Indiana. It is opening its bomb bay doors prior to releasing its single 5,000 pound bomb wrapped with multiple layers radioactive silica (sand). The radiation-spreading device was expected to glide forward and down self detonating over New York City's five boroughs spreading radiation and leading to radiation sickness and death among its several million inhabitants. Digital image by *Mario Merino*.

The 5,000 pound radioactive silica wrapped high explosive bomb is seen having just barely cleared the *Sänger* piloted, winged sub-orbital reusable rocket bomber to begins its downward glide eastward toward New York City. Digital image by *Mario Merino*.

The *Sänger* winged sub-orbital reusable rocket bomber, more easily maintains its altitude when it is minus its 5,000 pound radiation wrapped high explosive bomb. It now begins its homeward journey back to a wheeled landing site in *Nazi* Germany. Once on the ground its ground crew could make it ready for another mission in a few hours compared to two weeks (working 24 hours seven days a week) for the United States "Space Shuttle." Digital image by *Mario Merino*.

putting out 100-tons (200,000 pounds) of thrust. By attaching a one-man pressurized cockpit to a 100-foot long fuselage having short stubby wings and a tail assembly with dual rudders, *Sänger* hoped to create the first manned intercontinental missile-bomber capable of striking North America. Cities slated to receive its 1- to 5-ton payload (2,000 to 10,000 pounds) of high explosives, dropped from 30 to 100 miles up in space, included New York, NY (target for international banking and the American stock market); Washington, D.C. (target for U.S. national government and center for the war effort); Chicago, IL (target for much heavy industry and a transportation center); and Pittsburgh, PA (target for U.S. steelmaking activity). *Sänger's* orbital bomber was to have started its journey from a launch site somewhere within Germany — a third of the way around the globe from New York City. This is the reason the Soviets wanted to find *Sänger* so badly. Now embroiled in a Cold War with the West, they wanted the man who could design and build a machine capable of bombing North America from space.

Sänger's Amerika Bomber proposal had been known by various names since the early 1940s, when he first proposed the concept to the German *Luftwaffe*. Before that (1931-1932), *Sänger's* original thinking was to create a rocket-powered civilian aerospace transport flying machine. But when he presented his ideas for a rocket-powered flying machine capable of attacking North America — especially New York City or Washington, D.C. – to *Hermann Göring*, *Sänger* referred to his rocket-powered, ultra high altitude flying machine in war terms. Today, these names often include:

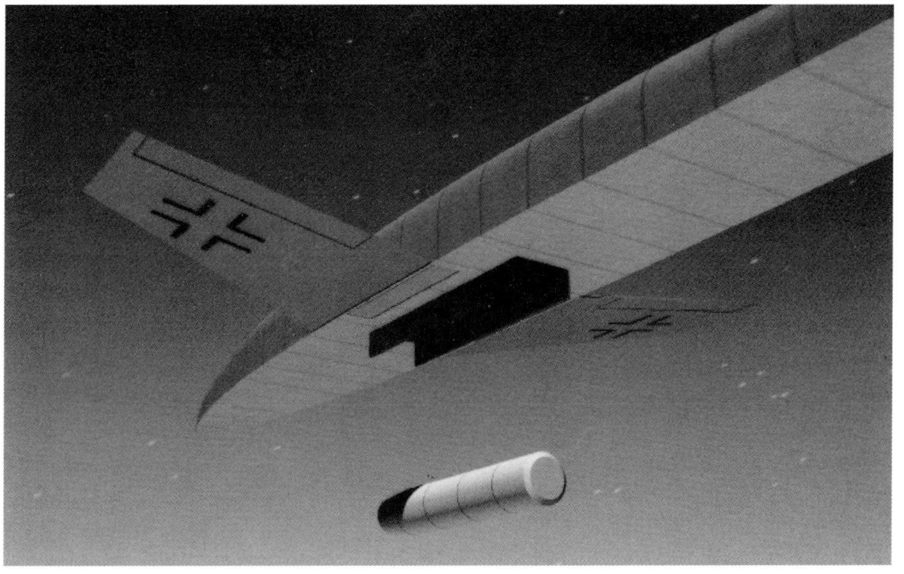

***Nazi* Germany's radiation-spreading bomb, now fully clear of the *Sänger*. It is only moments away from appearing over New York City since it, too, once had a forward speed of nearly 14,000 miles per hour before being released. Its forward speed would steadily slow due to atmospheric friction so that it would end its flight about 3,000 feet above metropolitan New York. Digital image by *Mario Merino*.**

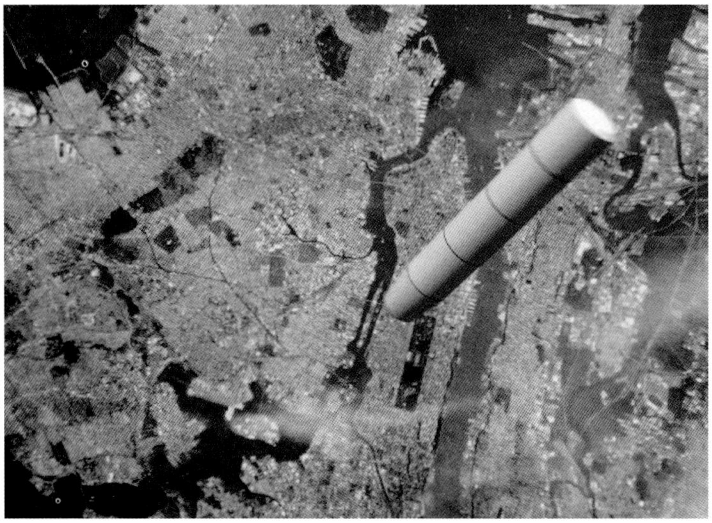

The nose of *Nazi* Germany's 5,000 pound bomb is pointing to New York's Manhattan Island directly beneath. The dark rectangle is Central Park. Digital image by *Mario Merino*.

All that New Yorkers see from the exploding *Luftwaffe* bomb three thousand feet above them is a flash of light followed by a rapidly spreading outward ring of fire carrying the deadly radioactive silica. Digital image by *Mario Merino*.

Glide Bomber, *Boost Glide Bomber*, *Skip Glide Bomber*, *Orbital Bomber*, and the *Antipodal Bomber*. The last two names are most commonly used to describe *Sänger's* 100-ton loaded weight space missile/bomber. The term "*Antipodal Bomber*" comes from a concept that "antipodes" are simply two points on the Earth exactly on opposite sides of the globe from one another.

Eugen Sänger strongly believed that the range of flight speeds several times the velocity of sound was the exclusive province of the pure rocket, which developed its propulsive thrust entirely from the fuel it carried on board the vehicle. The pure rocket also could compete in cost at lower speeds, if propulsive forces of great magnitude or short duration were required, or if no surrounding air was available – under water, for example, or outside the perceptible atmosphere of the earth. These were the special characteristics which gave rocket propulsion a unique domain in military techniques, including:

- Propulsion of projectiles or bombs in which the relatively stronger short duration propulsive forces could be achieved in most cases by powder-rockets;
- An auxiliary drive for propeller or jet aircraft with operative periods generally under a minute for which liquid rockets with compression-drive could be used;

The radioactive ring of fire is shown covering most of New York City's five boroughs as it keeps expanding outward, eventually covering neighboring New Jersey, too. Digital image by *Mario Merino*.

The Statue of Liberty can only watch in sadness as radioactive silica drifts down over her like flakes of snow in winter time. Digital image by *Mario Merino*.

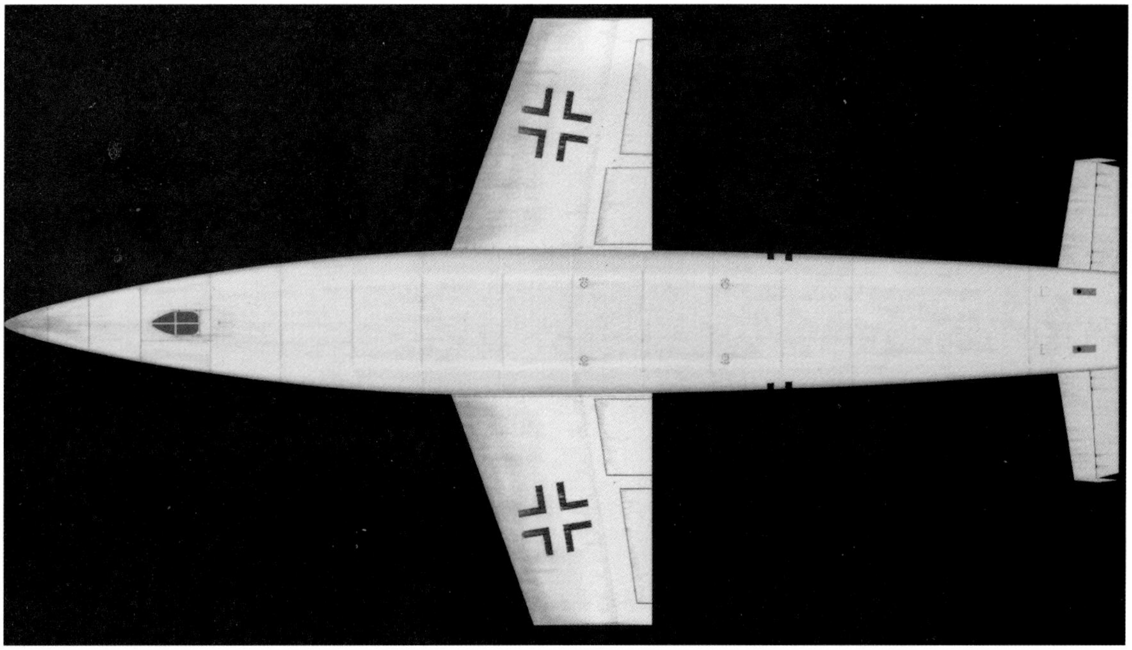

A top planform view of the *Sänger* piloted, winged, sub-orbital reusable rocket bomber showing its wing and tail control surfaces. The cockpit cabin design had not yet been settled on when *Eugen Sänger* proposed this offensive weapon to the *RLM*. It is featured here, however, with its pressurized single-seat cockpit with its windows flush with the surrounding fuselage. Digital image by *Josha Hildwine*.

- Auxiliary or principal propulsion for vessels with period of operation of several minutes using rocket motors having fuel pumps, but without a high exhaust speed;

- The main drive of aerial torpedoes against land, sea, or air targets with moderate-to-long duration of operation in which high exhaust speeds are important to achieve long ranges;

- The main drive of fighter or bomber aircraft — for fighter defense at very high altitudes or for military aircraft operating over very great distances. Both propulsion-time and exhaust speed set extreme requirements for the rocket motor;

Sänger stated that his proposed rocket bomber differed from the then present-day propeller-driven bomber aircraft in the following points:

- In place of the propeller propulsion from the front of the fuselage, it had rocket propulsion in the rear of the fuselage;

- The rocket aircraft's fuselage was in the shape of a bullet with a tapered afterbody;

- The wings had a thin wedge-shaped profile with sharp leading and trailing edges and a structure designed to withstand a high wing loading at the start of the flight;

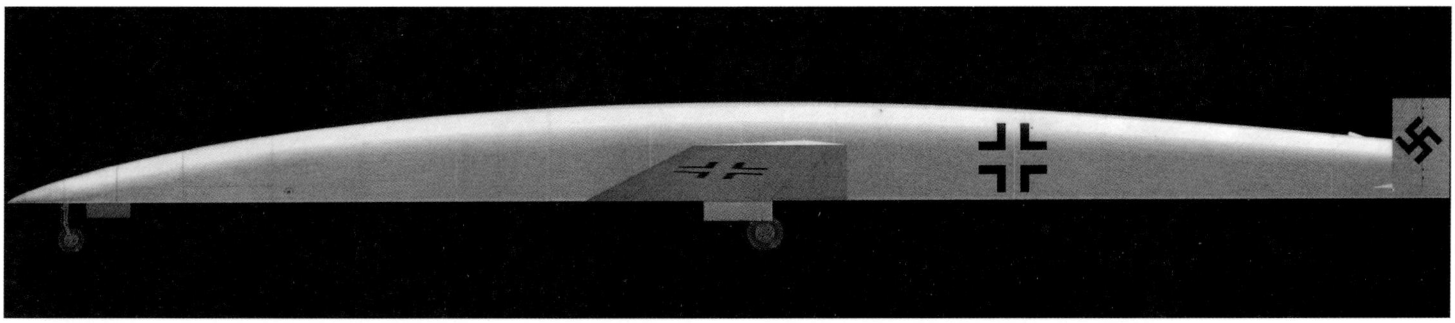

A port side view of the *Sänger* piloted, winged, sub-orbital reusable rocket bomber. The ultra high altitude flying machine is shown with a wheeled tricycle landing gear arrangement but it really was designed to be a tail-dragger to help it slow down after it landed. It would have had a nose wheel as shown...just for purposes of preventing it falling on its nose during landing. This issue of a tail dragged had not been settled by late 1944, at least. Discussions were underway as to the merits of a wheeled landing gear as shown or one with metal skids. Notice that the under surface of the orbital bomber was flat...to provide lift in the upper atmosphere so that it could push itself off and up from the dense atmosphere as it continued to proceed around the earth. Digital image by *Josha Hildwine*.

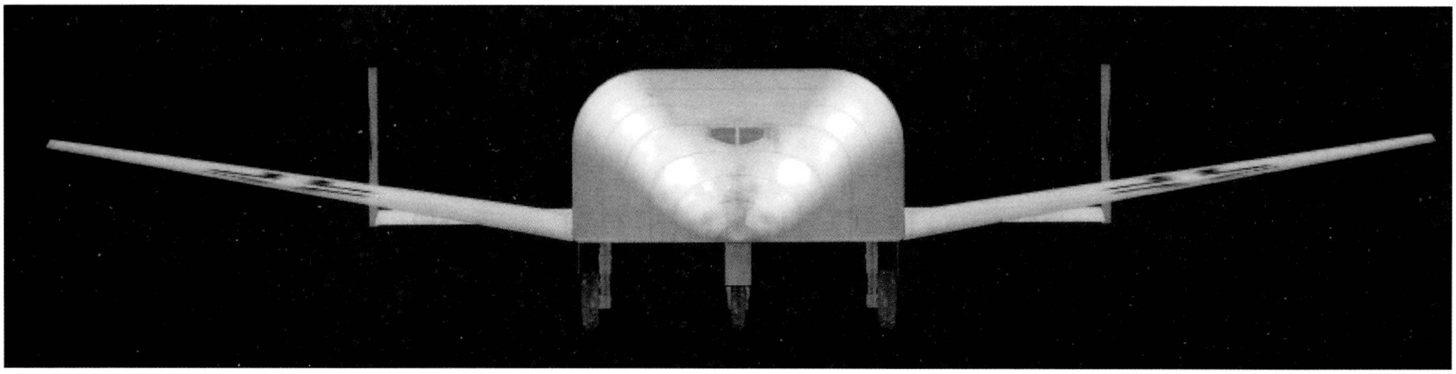

A nose-on view of *Eugen Sänger's* winged orbital reusable rocket bomber. This image features the wheeled tricycle landing gear and a fully enclosed, flush-mounted single seat cockpit cabin. Its tail skid cannot be seen from this angle. Digital image by *Josha Hildwine*.

- The pilot's cabin would be constructed as an airtight pressurized chamber capable of human life support outside most of the Earth's atmosphere;

From the beginning of his bomber studies, *Sänger* said his rocket bomber would use its own fuel just as in the usual propeller-driven aircraft. But because of the great difference in starting- and landing-weight it would lead to a large wing surface and too high fuel consumption in the range of speeds below the velocity of sound. A vehicle making a vertical takeoff under its own power, according to *Sänger*, had an even greater disadvantage and performance penalty. But push-starting a vehicle by using a rocket propelled sled on a horizontal takeoff monorail track until the velocity of sound had been reached, appeared to *Sänger* as a unique and logical solution. In this type of start by a captive rocket sled, the especially energy-hungry part of the aircraft's acceleration would not be obtained at the expense of fuel carried on board, so that the range of an aircraft catapulted in this fashion would extend significantly farther. At the same time, flight characteristics could be matched more easily to the steadily decreasing wing loading during flight. So as *Sänger* considered various methods of flight, he theorized that:

***Dr.-Ing. Eugen Sänger* (1905-1964) as seen in the early 1950s.**

- Acceleration to the point where flight speed equaled exhaust speed was possible and could be determined; and
- Acceleration to such a speed could be achieved so that the subsequent unpowered glide could extend over the entire remaining flight path;

Sänger argued that for equal fuel cost, the sled launching method used a simpler power plant and gave far greater range, so this became the basis of design for his rocket-bomber. The whole procedure would take place as follows: *Sänger's* rocket bomber, on the surface of the Earth and resting upon a monorail track, would be brought to a speed of about 500 meters/second [1,642 feet/second, or 1,075 miles/hour] by a launching sled — driven by captive rocket motors — running over a 3 kilometer [1.8 mile] long monorail track for a duration of 11 seconds. The bomber would then be released from the launching sled and climb, using its own internal rocket motor, up to an altitude of 50 to 150 kilometers [31 to 93 miles] along a path which is first inclined at 30 degrees to the horizon, but later flattens as peak altitude is reached. Finally, the vehicle reaches final velocities up to more than twice its climb rate. *Sänger* calculated that the duration of climb to peak altitude would be between 4 and 8 minutes and that during this climb period all the rocket bomber's fuel would be consumed. At the end of the climb, the rocket motor would be turned off and the rocket bomber, using its kinetic and potential energy, would continue on a path around the globe in a sort of oscillating gliding flight — or "skips" off the upper atmosphere — with a steadily decreasing amplitude of oscillation.

Sänger said that his rocket bomber's motion would be similar to the path of a long-range projectile which follows a descending glide-path. Yet, because of his rocket bomber's wings and flat fuselage underbody, the machine descending on its ballistic curve would bounce on the upper layers of the atmosphere and would again be kicked upwards, like a flat stone ricocheting off the surface of a still water pond. However, as the rocket bomber passed through denser air each time, a fraction of its kinetic energy would be consumed. The intensity of the initial rebounds off the atmosphere would steadily become smaller and the frequency of contact with air would increase, finally bringing the vehicle to a steady gliding flight. At the same time, its flight speed along the glide path of, say 2,000 kilometers [1,243 miles], would drop from an initially high launch speed to a more normal gliding and landing speed. If the descent path (which is within certain limits controllable by the pilot) corresponded to the direction of the target, the bombs would be released at a predetermined moment, and the *Sänger* rocket bomber would then return to its starting place. On the other hand, the rocket bomber could head for an alternative field or be sacrificed — with the pilot ejecting in his escape capsule. The launched bombs would continue to the target along the original direction of flight.

Sänger believed that even if the target was very distant from his rocket bomber's take off point, the bombs needed only be dropped near the target which would then be covered by a *Gaussian* distribution of hits. This military use of the *Sänger* rocket vehicle would be the first 'space bomber' – flying completely unaffected by time of day, weather conditions at the target, and any enemy air defenses, because of the vehicle's great altitude and speed.

From the characteristics given by *Sänger* about his rocket bomber, it appears that his concept was not merely a development of an improved military flying machine which could replace existing types. It was a quantum leap forward in weapons system theory. *Sänger* believed he had solved a problem for which no solution existed up to that time...namely providing a method for bombing targets at distances of 1,000 to 20,000 kilometers [620 to 12,400

Eindringtiefe deutscher Fernkampfflugzeuge, Frühjahr 1944

(Ju 390, Me 264 errechnete Werte) Quelle: BA / MA, RL8/187k, RL 36/119

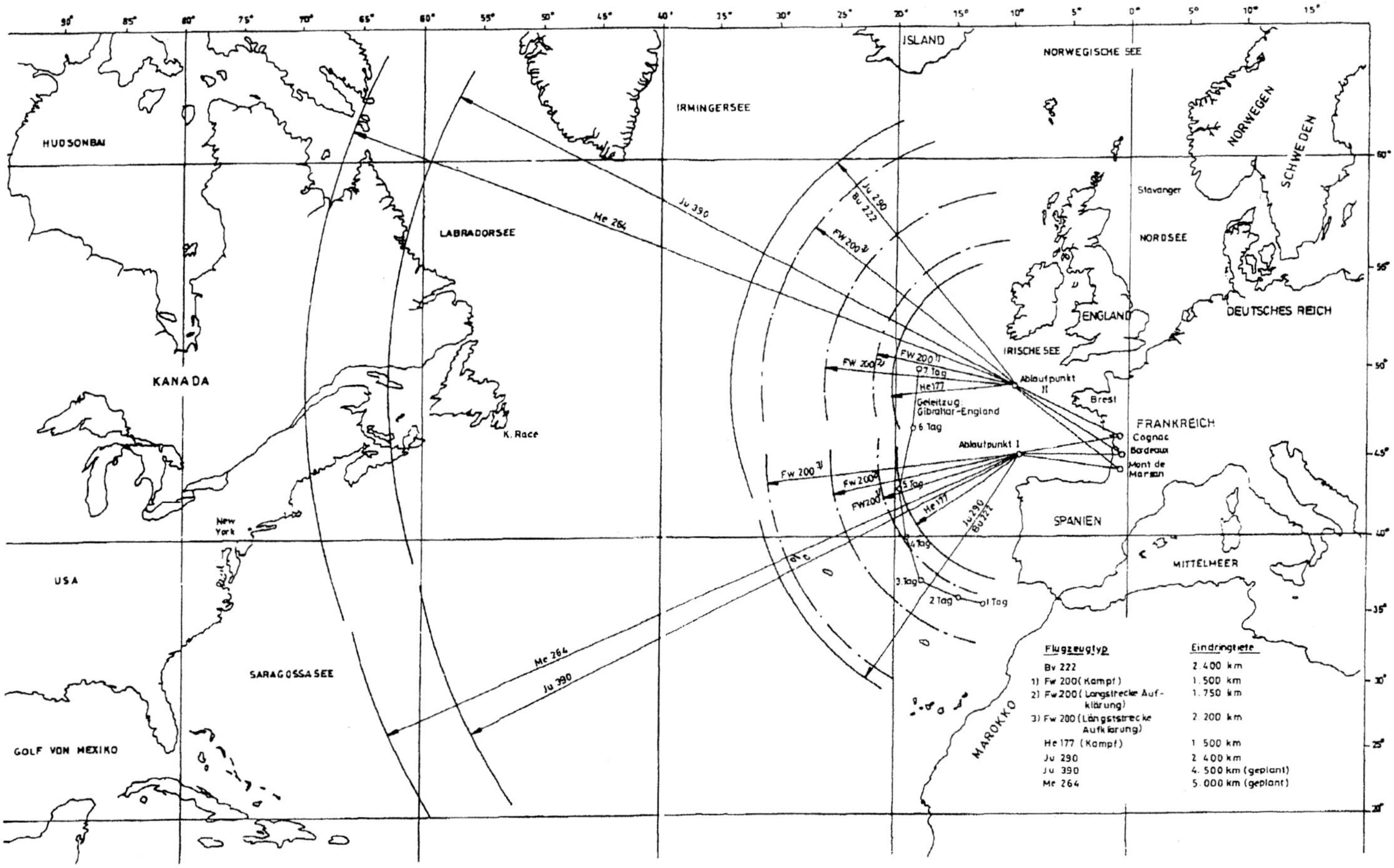

This *Nazi* Germany pen and ink map from 1944, features about how far existing *Luftwaffe* multi-piston engine heavy bombers could be expected to reach in their efforts to hit America after takeoff from an airbase in German occupied France. *Junkers Ju 390s* would barely reach the coast of Canada. A six-motored *Messerschmitt Me 264* might reach a little deeper in to Canada and the probably State of Maine. So neither the *Junkers Ju 390* or the *Messerschmitt Me 264* could be expected to even reach New York City...let alone return home. *Sänger* told the *RLM* that his piloted winged sub-orbital rocket bomber could reach any city in the United States and easily return home to its launch site in *Nazi* Germany.

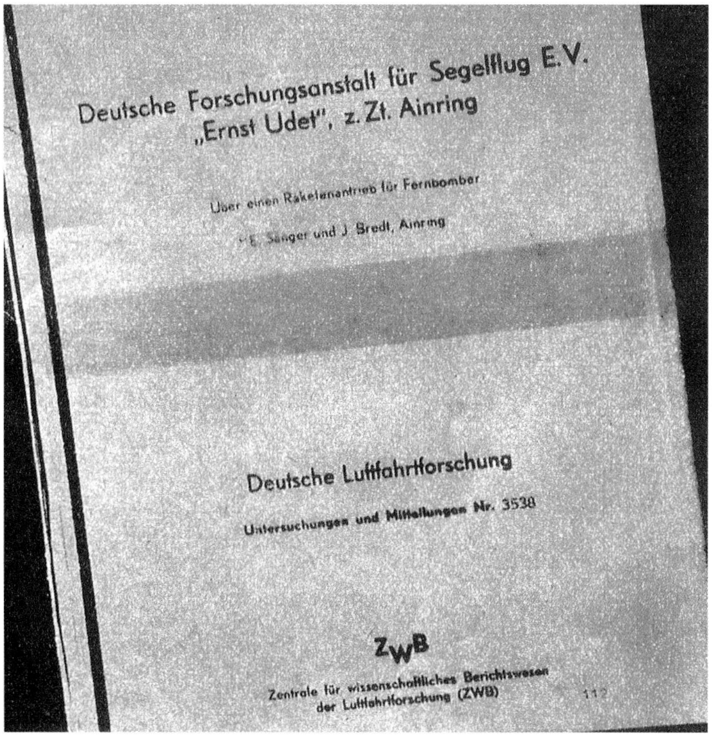
Deutsche Forschungsanstalt für Segelflug E.V.
„Ernst Udet", z. Zt. Ainring

Über einen Raketenantrieb für Fernbomber

E. Sänger und J. Bredt, Ainring

Deutsche Luftfahrtforschung

Untersuchungen und Mitteilungen Nr. 3538

ZWB

Zentrale für wissenschaftliches Berichtswesen der Luftfahrtforschung (ZWB)

112

The cover of *Eugen Sänger* and *Irene Bredt's* August 1944 proposal on a rocket drive for long range bombers. This particular copy is copy #112. It is believed that only 125 copies of the highly secret document was published. *Eugen Sänger's* boss at *DFS*, *Professor Walter Georgii*, was instrumental in getting *Sänger's* controversial ideas printed and passed around to various members of Germany's aviation community and rocketeers.

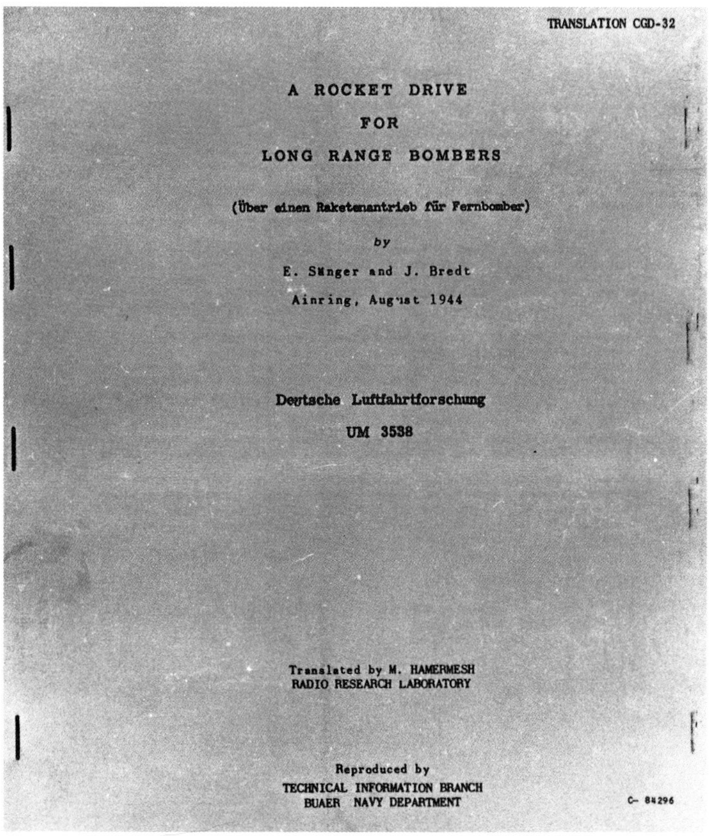
TRANSLATION CGD-32

A ROCKET DRIVE
FOR
LONG RANGE BOMBERS

(Über einen Raketenantrieb für Fernbomber)

by

E. Sänger and J. Bredt

Ainring, August 1944

Deutsche Luftfahrtforschung

UM 3538

Translated by M. HAMERMESH
RADIO RESEARCH LABORATORY

Reproduced by
TECHNICAL INFORMATION BRANCH
BUAER NAVY DEPARTMENT

C- 84296

miles] from its take-off point. The *Sänger* rocket bomber theoretically would be so efficient that now single point bombing attacks could be made from Central Europe on distant targets anywhere on the globe...New York, Moscow, London as well as the Panama Canal. With a group of only 100 relaunchable rocket bombers, according to *Sänger,* areas the size of a large city located anyplace on earth could be completely destroyed within a few days' time.

RAKETENBOMBER Specifications:

- **Aircraft type -** Monorail launched winged piloted reusable global range rocket bomber**;**
- **Armament -** None;
- **Engine -** 100-ton (200,000 pounds) thrust bi-fuel liquid rocket engine utilizing liquid oxygen, gasoline, and diesel oil;
- **Engines, booster sled** - *Sänger* claimed only two *Wernher von Braun V2* type bi-fuel liquid rocket engines of 50,000 pounds thrust each would be needed with a burn time of approximately 11 seconds. *Al Bowers* of NASA-Dryden calculates that as many as 30 *V2* engines might have been required to push the 100-ton *Sänger* orbital rocket bomber and its launch sled to the end of its 1.86 mile long rail to reach a speed of 1,113 miles/hour;
- **Disposition -** A proposal only. No detailed drawings were ever started;
- **Summary** - Though the *Sänger* liquid rocket propelled bomber never even got into the preliminary design stage, a 100-ton thrust prototype combustion chamber was under construction at war's end. Research work in high temperature alloy construction, orbital mechanics, exo-atmospheric 3-axis control, inertial navigation, re-entry dynamics, and advanced sensor systems also was underway at the time of war's end in May, 1945. *Sänger's* orbital bomber idea was revisited in the early 1950s by *Bell Aero-Systems* (a division of *Bell Aircraft Corporation*) in a proposal for the U.S. Air Force for a rocket-launched global range bomber. This project was under the direction of *Herr Krafft Ebhicke* and *Dr.-Ing. Walter Dornberger* (1895-1980), who was director at the former German secret weapons development center at Peenemünde. It was known as the Bell *"Bomi"* (*Bomi*, being the acronym for "bomber-missile"). The *Bomi* was to have been a glide-vehicle boosted by a two-stage rocket to altitudes of 100,000 feet or higher, at speeds of *Mach 4*. Range was to have been at least 3,000 miles. The proposed *Bomi* by *Bell Aero-Systems* received an unfavorable evaluation by the Air Force in 1954. In 1957, a proposal issued by the USAF's Air Research and Development Command called for an orbital re-entry glide vehicle to be called *Dyna-Soar* (Dynamic Soaring) and later called the *X-20*. In November, 1959, the Air Force selected *Boeing Aircraft* as the sys-

Left: *General George McDonald's* USAAF Intelligence people found a copy of the *Sänger-Bredt* proposal at *DFS*-Ainring about the time of Germany's May 1945 unconditional surrender. *General McDonald* had it immediately translated into English, although it was still classified "Secret."

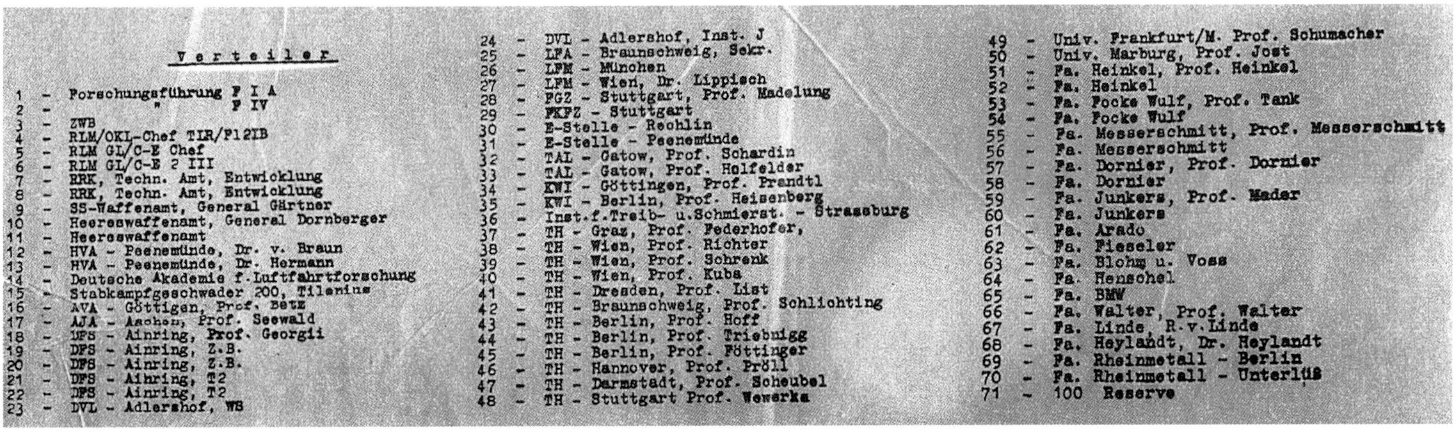

Verteiler

1 - Forschungsführung F I A
2 - " F IV
3 - ZWB
4 - RLM/OKL-Chef TLR/F12IB
5 - RLM GL/C-E Chef
6 - RLM GL/C-E 2 III
7 - RRK, Techn. Amt, Entwicklung
8 - RRK, Techn. Amt, Entwicklung
9 - SS-Waffenamt, General Gärtner
10 - Heereswaffenamt, General Dornberger
11 - Heereswaffenamt
12 - HVA - Peenemünde, Dr. v. Braun
13 - HVA - Peenemünde, Dr. Hermann
14 - Deutsche Akademie f. Luftfahrtforschung
15 - Stabkampfgeschwader 200, Tilenius
16 - AVA - Göttigen, Prof. Betz
17 - AJA - Aachen, Prof. Seewald
18 - DFS - Ainring, Prof. Georgii
19 - DFS - Ainring, Z.B.
20 - DFS - Ainring, Z.B.
21 - DFS - Aihring, T2
22 - DFS - Ainring, T2
23 - DVL - Adlershof, WB
24 - DVL - Adlershof, Inst. J
25 - LFA - Braunschweig, Sekr.
26 - LFM - München
27 - LFM - Wien, Dr. Lippisch
28 - FGZ - Stuttgart, Prof. Madelung
29 - FKFZ - Stuttgart
30 - E-Stelle - Rechlin
31 - E-Stelle - Peenemünde
32 - TAL - Gatow, Prof. Schardin
33 - TAL - Gatow, Prof. Holfelder
34 - KWI - Göttingen, Prof. Prandtl
35 - KWI - Berlin, Prof. Heisenberg
36 - Inst.f.Treib- u.Schmierst. - Strassburg
37 - TH - Graz, Prof. Federhofer,
38 - TH - Wien, Prof. Richter
39 - TH - Wien, Prof. Schrenk
40 - TH - Wien, Prof. Kuba
41 - TH - Dresden, Prof. List
42 - TH - Braunschweig, Prof. Schlichting
43 - TH - Berlin, Prof. Hoff
44 - TH - Berlin, Prof. Triebnigg
45 - TH - Berlin, Prof. Föttinger
46 - TH - Hannover, Prof. Pröll
47 - TH - Darmstadt, Prof. Scheubel
48 - TH - Stuttgart Prof. Wewerka
49 - Univ. Frankfurt/M. Prof. Schumacher
50 - Univ. Marburg, Prof. Jost
51 - Fa. Heinkel, Prof. Heinkel
52 - Fa. Heinkel
53 - Fa. Focke Wulf, Prof. Tank
54 - Fa. Focke Wulf
55 - Fa. Messerschmitt, Prof. Messerschmitt
56 - Fa. Messerschmitt
57 - Fa. Dornier, Prof. Dornier
58 - Fa. Dornier
59 - Fa. Junkers, Prof. Mader
60 - Fa. Junkers
61 - Fa. Arado
62 - Fa. Fieseler
63 - Fa. Blohm u. Voss
64 - Fa. Henschel
65 - Fa. BMW
66 - Fa. Walter, Prof. Walter
67 - Fa. Linde, R.v.Linde
68 - Fa. Heylandt, Dr. Heylandt
69 - Fa. Rheinmetall - Berlin
70 - Fa. Rheinmetall - Unterlüß
71 - 100 Reserve

This is a list of individuals, aviation companies, and research centers receiving a copy of the original *Sänger-Bredt* proposal in August 1944. It appears that 70 copies were dispersed and 55 copies held in reserve. This list reads like a who's who of *Nazi* Germany's aviation community, aviation researchers, and rocketeers.

tem contractor and *Martin Aircraft* Company as associate booster contractor. A contract was issued in April, 1960. *Bell-Aero Systems* dropped out of the orbital vehicle program entirely.

The *Sänger* Sub-Orbital Rocket Bomber Overall Specifications Include:

- **Wingspan** - 15.0 meters [49.2 feet];
- **Length, overall** - 28.0 meters [91.8 feet];
- **Height** - Not available;
- **Diameter, core** - [7 feet 5 inches];
- **Weight, aircraft empty** - approximately [18,000 pounds] or 9-ton;
- **Weight, gross, at launch** - approximately [200,000 to 208,000 pounds] or 100- to 104-ton;
- **Weight, fuel -** approximately [180,000 pounds] or 90-ton;
- **Weight, bomb load -** 1- to 5-tons [2,000 to 10,000 pounds];
- **Speed, maximum** - 21,880 kilometers/hour [13,596 miles/hour] at suborbit insertion;
- **Speed, cruise** - 20,921 kilometers/hour [13,000 miles/hour] initially, then declining;
- **Landing mechanism** - A skid-type 3-point landing gear consisting of three metal alloy skids (wheels were also mentioned), mated to retractable struts having an air/oil shock absorber that would minimize drag after surface contact. A nose wheel would be used only to prevent the machine from bouncing down on its nose upon landing, otherwise the vehicle would make its landing approach in a nose-high attitude to slow it down. No drag parachute was mentioned;
- **Speed, landing** - An approach speed of about 402 kilometers/hour [250 miles/hour], with a landing speed of roughly 258 kilometers/hour [160 miles/hour];
- **Launching** - launched via multiple captive rocket booster units on a sled riding a horizontal monorail track approximately 1.86 miles [3.5 kilometers] in length;
- **Lift off angle after launching** - 30 degrees up to 500,000 feet (94.7 miles) altitude;
- **Speed, lift off** - *Mach* 1.5 or 1,113 miles/hour; *

* [Note] The speed of sound varies according to altitude and temperature. For example, the speed of sound is 742 miles per hour at sea level. It decreases with altitude until it reaches 661 miles per hour at 36,000 feet, then remains at that speed in a band of steady temperature up to 60,000 feet altitude.

The *Sänger* piloted, winged, sub-orbital, reusable rocket bomber appears waiting for a mission as seen looking down on its longitudinal axis and from its starboard side. Digital image by *Mario Merino*

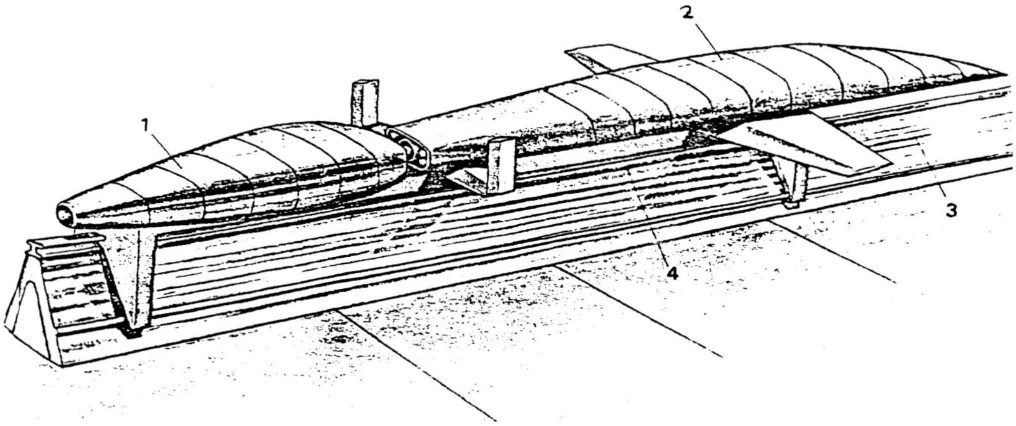

A copy of the original pen and ink drawing by *Sänger* featuring the proposed horizontal launching arrangement (1942-1944) for his piloted, winged, sub-orbital, reusable rocket bomber. 01 - **Bi-fuel liquid captive rocket powered booster sled;** 02 - **The 91 3/4 foot long *Sänger* bi-fuel rocket powered winged orbital reusable bomber;** 03 - **Triangular concrete structure of the 1.86 mile-long monorail launch track;** 04 - **Steel track supporting both the 100-ton winged bomber and its heavy rocket booster sled with its several captive *Wernher von Braun A-4 (V2)* 50,000 pound thrust rocket engines and fuel.**

- **Speed, post-launch at 500,000 Feet [94.7 miles] altitude** - Upon leaving its horizontal take off monorail, the rocket bomber's internal 100-ton thrust bi-fuel liquid rocket engine would be ignited and burn for approximately 4 to 8 minutes, consuming all of its 90-tons of fuel and reaching an altitude of 1,524 kilometers [94.7 miles], where it would reach a speed of approximately 21,880 kilometers/hour [13,596 miles/hour].
- **Launch time to 500,000 feet altitude** - *Sänger* claimed 4 to 8 minutes. *Al Bowers*, an NASA aeronautical engineer, calculates it would have reached its operational altitude between 7 and 8 minutes.
- **Range, maximum** - 23,489 kilometers [14,596 miles];
- **Altitude, maximum** - approximately 152 kilometers [95 miles];
- **Operational cruise altitude (initial)** - 145 kilometers [90 miles];
- **Crew**- 1 seated in conventional seating in a cabin pressurized for attitude;
- **Crew emergency escape** - Jettisonable cockpit capsule designed by *DFS*, similar in operation to capsules designed for the *DFS 228V3* and *DFS 346* aircraft;
- **Airframe** - External shell to be constructed of metal alloys capable of maintaining rigidity and strength at extremely high temperatures and aerodynamic loads;
- **Fuselage** - Ballistic-shaped with flattened underside containing cockpit cabin, two oxygen tanks – one each side of the fuselage forward of the wing; also containing a 100-ton thrust bi-fuel rocket engine with required fuel and oxidizer tanks located in the aft fuselage and flanked by two auxiliary rocket engines;
- **Wing structure** - Short triangular wedge type tapered airfoil section with a maximum thickness of 1/20th of the chord 2/3 back from leading edge;

This is an artist's rendition featuring the *Sänger* piloted, winged, sub-orbital, reusable, rocket bomber with its captive booster as it might have looked atop its 1.86 mile long monorail launch as it prepared for takeoff. Digital image by *Josha Hildwine*.

Left: *Eugen Sänger,* in the early 1950's, has his right hand finger on a globe explaining how his piloted, winged sub-orbital, reusable, rocket bomber design of 1944-1945. This flying machine would have sort of skipped along its upper atmosphere on its way to North America, especially to the United States and her cities of New York and Washington, D.C.

- **Tail assembly** - A horizontal tail assembly with small fins on each side with attached hinged rudders;
- **Color, exterior** - Silver overall;
- **Cooling, exterior** - Radiation cooling through the use of a water wall that utilized latent heat of evaporation for heat dissipation during atmospheric re-entry from sub-orbital missions;
- **Cooling, interior** - Cockpit air-conditioning and pressurization were to be furnished by a ram-air system;
- **Control, post re-entry** - Conventional full 3-axis control with trailing edge ailerons, elevators, and dual vertical rudders, hydraulically actuated through a center stick, to control the aircraft during low- speed atomospheric flight;
- **Control, ex-atmosphere** - Four small liquid rockets located aft of the main combustion chamber.

The *Sänger* piloted, winged, sub-orbital reusable rocket bomber shown cruising about 80 miles above the earth in the vicinity of the Great Lakes (Chicago, Illinois) with a forward speed of nearly 14,000 miles per hour. Digital image by *Dan Johnson*.

Characteristics of the Sänger Rocket Bomber

Eugen Sänger's proposed manned rocket bomber had a unique appearance unlike any flying machine seen in Germany up to that time. The aircraft's upper fuselage was "ogival", with a 9.6 radius of curvature. Between the wings, this semi-ogive transitioned into a long chamber with perpendicular side walls. The fuselage gradually tapered toward the stern with a steady decrease in cross-section. The large blunt end surface at the fuselage aft end was necessitated by the size of the rocket motor's combustion chamber. The relatively small, stubby wings would have served mainly for stabilization in flight and for landing. The wing cross-section had a triangular "wedge" profile, with a maximum thickness of 1/20 of the chord length, located at a distance two thirds behind the leading edge. *Sänger* claimed that this particular aircraft fuselage and wing shape corresponded to the laws of flow for very high *Mach* numbers. He believed that a positive angle of incidence between the fuselage and wings was unnecessary for the low-wing arrangement chosen. The "lifting" flat surfaces of the fuselage underbody and both wings blend into one another without a break. The tail surface on the rocket bomber was a simple symmetrical quadrangular cross-section. *Sänger* said that its greatest thickness was 1/20 of its depth in the last 3rd of its chord. The entire tail assembly was not affected by the exhaust from the rocket combustion chamber since use of the rocket motor and flight below sound velocity would never occur together.

The rocket bomber would have been a very large flying machine. *Sänger* said that the size of his rocket bomber was chosen as a compromise between a series of contradictory requirements. The idea of making the aircraft as large as possible is suggested by the fact that a ratio of higher payload capacity to weight when empty is generally more favorable in an aircraft. This means, too, that the construction of larger rocket motors becomes simpler. With increased aircraft size and payload, the military strength of a rocket bomber group increases, because the number of capable pilots required per unit of weapons payload transported decreases. *Sänger* claimed that if one computed a few comparative designs in the range

A United States Air Force navigator sits on the floor of a transport aircraft catching up on the latest news: "*Hitler Dead.*" He had been shot down over Germany several weeks earlier, rescued, and was being taken back to his air base in England. With *Hitler's* death, *Eugen Sänger* wondered to whom should he transfer his allegiance...America, England, France, or the Soviet Union? Or did this news of *Hitler's* death also mean the end of *Sänger's* winged orbital bomber? Of the several Allies, *Eugen Sänger* chose France.

The *Sänger-Bredt, Nazi* Germany's piloted, winged, sub-orbital, reusable rocket bomber dressed up post war with the Soviet red star on its wings and fuselage side. Courtesy: *John W.R. Taylor*, "*Russia's Secret Bomber*?" *Flying*, October 1953.

of 10- to 100-tons starting weight, one found that with increasing vehicle weight, the aerodynamic lifting contributed by the fuselage represents (for geometric reasons) a smaller part of the total weight, so that the wings have to be relatively larger. However, the weight of larger wings will predominate without giving any noticeable improvement in gliding angle in the region of high *Mach* numbers. Such considerations by *Sänger* led to a 'formula' for a takeoff weight of 100-tons and an empty weight on landing of 10-tons. Thus a limit of 90-tons of fuel contained within 76 cubic meters of tank space was derived. These assumptions led to the rocket bomber's huge overall dimensions.

Wing dimensions proposed by *Sänger* were based on the rocket bomber's projected wing loading. Since the aircraft's starting procedure would have used a launch sled with captive *Wernher von Braun V2* rocket motors, a high wing loading would have been permissible at launch. Assuming the consumption of all fuel and removal of all weapons 'ballast' would have reduced the rocket bomber's weight to 1/10 of its gross takeoff weight, its landing speed still was the major determinant of the wing size. *Sänger* figured that landing speeds of over 250 kilometers/hour [160 miles/hour] might occur in special cases, a more real-life limit of 200 kilometers/hour [125 miles/hour] was chosen because the landing of his rocket bomber was to be considered a "glide" landing. Upon touchdown it would fall back on its tail skid to help slow down its forward motion.

***Hans Motsch* in the mid 1940s. He was a former test pilot at *Siebel-Halle* and post war was transferred to the USSR to continue constructing the *DFS 346* and *Junkers Ju 287 V3* prototypes. For some unexplained reason, young *Stalin* seemed convinced that *Motsch* knew the where *Sänger* and *Bredt* could be found post war and was friendly. *Hans Motsch* told this author in the mid 1980s during an interview in the *DDR* that if he had known that the *Sänger's* were living in Paris, France, then he would have become free and a very rich man, too.**

An image of *Vasilli Stalin*...highly retouched by Soviet photographers of the late 1940s, as was their custom. Young *Stalin* was authorized by his father to give a person's weight in gold if they could provide the country and city where *Eugen Sänger* and *Irene Bredt* could be found.

The very thin and slightly curved wing profile was expected to provide a maximum lift coefficient only equal to 1.25 kilograms per square meter – even with landing aids extended, so *Sänger's* rocket bomber's wing loading before landing would have been limited to 137.5 kilograms per square meter. *Sänger* said that aside from this figure, the rocket bomber's wing size was determined by the fact that the fuselage underbody plays an important role in the lifting power of the whole aircraft. According to his investigations for high *Mach* numbers, 2/3 of the machine's total weight would be carried by the fuselage and 1/3 by its wings. At landing speed, *Sänger* believed that the lift coefficient of the fuselage, at the best angle of attack for maximum lift, would be Cl = 0.45 and with the already

Two former *Siebel* test pilots: *Hans Motsch* (left) and *Barnard Trauter* (right), Bitterfeld, East Germany. Both were forced to the USSR post war and were not returned until 1954. Photo by author in the mid 1980s.

fixed fuselage supporting surface Fr. Ca. q = 4,000 kilograms, while the residual landing weight of 6,000 kilograms [13,228 pounds] went to its wings. Thus *Sänger* claimed that the required wing size for his rocket bomber was about 44 meters^2, the total supporting surface was 125.5 meters^2, the mean wing loading was 10,000/125.5 = 79.7 kilograms/meter^2, and the lift coefficient for landing was Cl = A/qF = 0.74.

Wing loading at takeoff, *Sänger* estimated, would be ten times as high...797 kilograms/meters^2 for the assumed climbing speed of 500 meters per second [1,642 feet per second]. At a 3 degree angle of attack, he estimated that the 38 percent of the total weight of 105 kilograms fell on the fuselage and 62 percent on its wings and the wing loading was 1,390 kilograms/meters^2. After the aircraft rose to the optimum angle of attack of 8 degrees, Ca became 0.173; thus lift for hauling the aircraft at V/a = 1.5 was 346,000 kilograms, of which 48 percent rested on its wings with CaF = 0.088, and corresponding to a wing loading of 3,720 kilograms/meters_ in this state of flight which determined the wing stress.

Sänger realized that other items/parts in his proposed rocket bomber would face heavy load forces. These included the loaded fuel tanks during takeoff, the fuselage and pilot and his acceleration until the end of the machine's climb, and the fuselage and its landing gear when landing.

In estimating the ratio of empty-weight G, after consumption of all fuel and dropping the useful load, to the flight weight Go immediately after rising from the ground, *Sänger* started from the known weight distribution for overloaded long range aircraft: airframe 18%, power plant 13%, auxiliaries 3%, additional load 66%, so Go = 0.34. The main parts of the rocket bomber and their individual weight was estimated by *Sänger* as:

- **Cockpit Cabin** - 500 kilograms [1,102 pounds];
- **100-ton bi-liquid rocket motor** - 2,500 kilograms [5,512 pounds];
- **Wings** - 2,500 kilograms [5,512 pounds] or a total weight of 56 kilograms/meter^2;
- **Fuselage** - 3,250 kilograms [7,165 pounds];
- **Tail assembly, landing gear including retractable tail skid, and bomb bay** - 1,250 kilograms [2,756 pounds];

Josep Stalin. He wanted _Sänger_ and _Bredt_ so badly, even if it took millions in reward money/gold leading to their capture and forced return to the USSR. In the case of _Sänger_ and his plans for a sub-orbital bomber, which could bomb America from space...money/gold was no issue.

A granite block statue of *Josep Stalin* near where the interned German aviation specialists, such as *Hans Motsch* and *Barnard Trauter*, labored (1946-1954) in the former USSR.

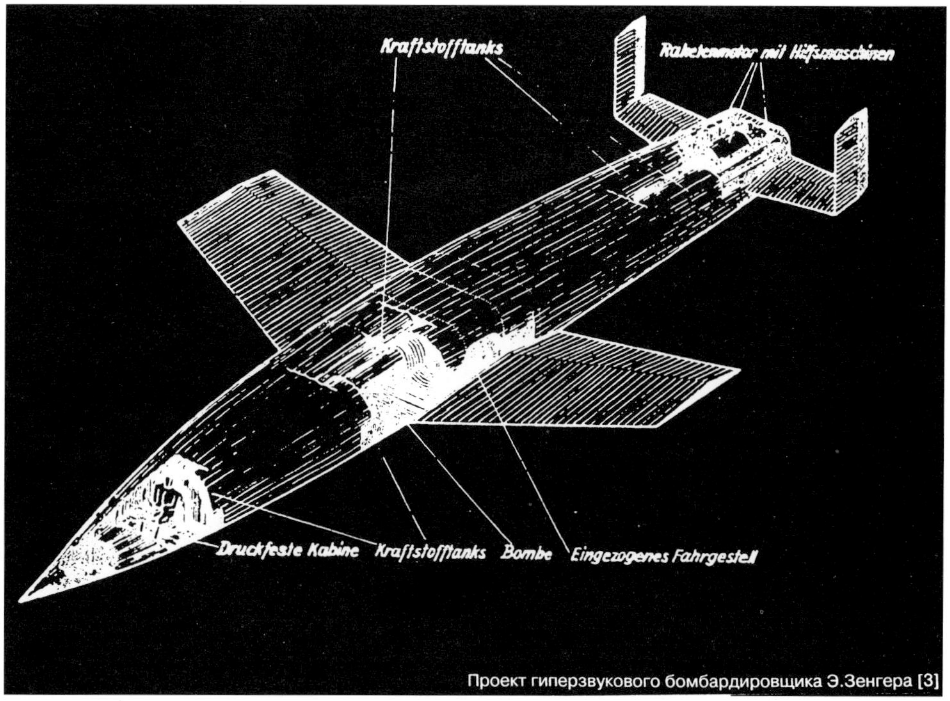

The Soviets translated the secret 1944 *Sänger-Bredt* piloted, winged, sub-orbital, reusable rocket bomber proposal. Shown is a pen and ink drawing from this document in German, however, with a Russian language caption at the bottom.

Thus, the total estimated empty weight of *Sänger's* rocket bomber would have been 10,000 kilograms [15,432 pounds]. *Sänger* said that the flight weights for ordinary bomber of similar size was about 3,000 kilograms [6,614 pounds]. The reason that his rocket bomber with its 10 times greater takeoff weight was only 3.3 times as heavy as an ordinary bomber of the same size was mainly because the supporting surfaces, especially the wings, would carry not 10 times, but only 3-4 times the weight, while the remainder was directly supported by the air against the fuselage without any intermediary structure. Moreover, because of the different method used for starting and climbing, the factor of safety for the rocket bomber need be only a small fraction of that for an ordinary bomber. Thus, the total weight distribution of the rocket bomber's airframe was estimated to be:

- **Airframe** - 7 percent;
- **Rocket engine** - 2.5 percent;
- **Auxiliary equipment** - 0.5 percent;
- **Additional load** (fuel and weapon payload) - 90 percent;

Front views of *Sänger's* rocket bomber do not normally show its retractable nose wheel. However, it operated in conjunction with a retractable tail skid and the main landing gear struts, which also were retractable into the fuselage between the wings. This landing

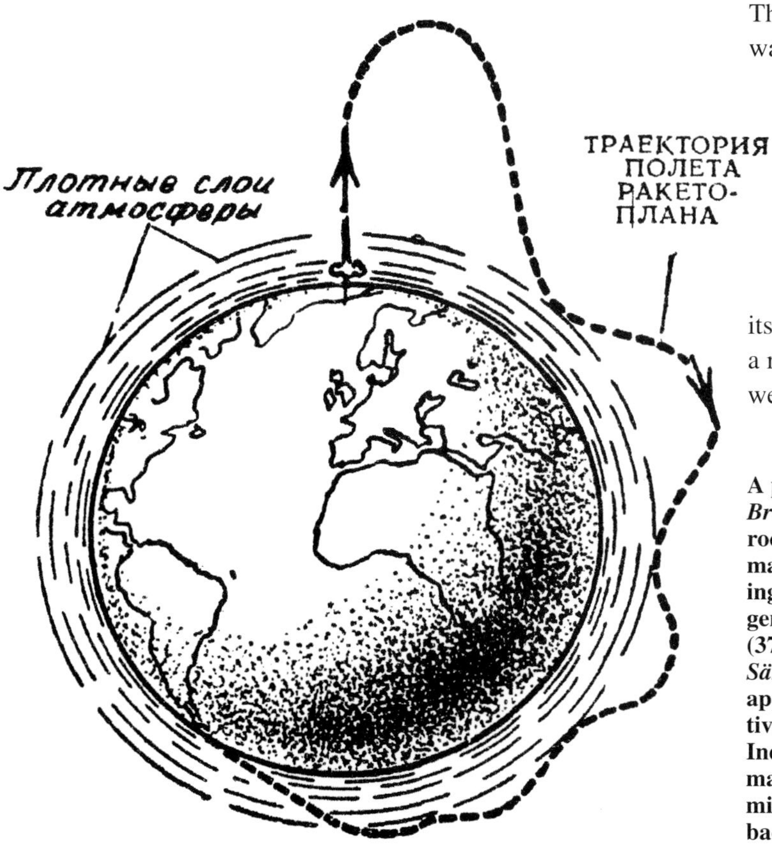

A pen and ink illustration from the Soviet translation of the *Sänger-Bredt* proposal featuring the route to be taken by the proposed orbital rocket bomber. After a steep angle of lift off it would leave *Nazi* Germany, heading east around the globe to attack New York City approaching from the Pacific Ocean. It appears that the Soviets changed the gentle undulating flight path of the orbital bomber, bringing it low (37 miles altitude) over its intended target: New York City. *Eugen Sänger* claimed that the lowest altitude over New York City would be approximately 78 miles. It really didn't, matter because the radioactive conventional bomb would have been released over Indianapolis, Indiana. Perhaps if the *Sänger* piloted, winged, rocket bomber had made a U-turn over New York City, then it might have dropped to 37 miles altitude, losing considerable height as it turned around to head back home to Germany.

Alleged Soviet spy and Soviet air force rocket scientist *Leutnant Colonel Grigori A. Tokaev*. *Josep Stalin* gave him the job of finding the *Sängers* assisted by fun-seeking son...*Vasilli Stalin*. The duo never found the *Sängers* although they later zeroed in on France. *Hans Motsch* told this author that *Vasilli* told him that he found Paris very much to his liking and began thinking less and less of the *Sängers*.

gear configuration is similar to that used on the North American *X-15* rocket-propelled research aircraft flown 15 years later by the National Aeronautics and Space Administration (NASA) and the United States. Air Force. The nose wheel served only to prevent dangerous nose contact with the ground during the bouncing motion of the machine when touching down at roughly 125 miles per hour. *Sänger* said that his rocket bomber would have practically no wind resistance because of its small wings and highly streamlined 'lifting body' fuselage. Behind the nose was the pressure-tight cabin in which a single pilot would sit. It was to be pressurized at 0.4 to 0.5 atmospheres and even with total vacuum outside, this pressurization should have permitted rapid exit of the pilot's ejection capsule in case of a life-threatening danger. It also was *Sänger's* idea that the pilot's seat be arranged in the best possible position so that the pilot could absorb the heavy acceleration occurring along the machine's flight axis during the launch phase. *Sänger* believed that not only should a pilot's body and head be protected, but also his feet and arms should have good supporting surfaces and that provisions be made so that body positions could be shifted. He had benefitted by medical research/studies of *Dr.-Med. Hubert Strughold*, a *Luftwaffe* surgeon, who today is known the world over as the father of space flight medicine. *Strughold* had a research data base on human volunteers who had withstood forces up to 17 times those

An overhead view of the full scale *DFS 346* being tested in a Soviet wind tunnel post war.

of normal gravity. He had used drugged monkeys — spinning them in centrifuges for a minute or more and subjecting them to forces up to 20 times those of normal gravity. *Sänger* did not give many clues to proposed flight instruments, other than mere mention that the cabin would be equipped with D/F (directional finding) instruments and radio equipment.

Behind the pressure cabin were to be fuel and oxidizer tank installations which consisted of two large tubes 20.5 meters [67.3 feet] long and with a maximum diameter of 1.8 meters [5.9 feet]. These tanks would constitute the main part of the fuselage. The upper 4th of the tubes' circumference would form the skin of the aircraft, while the lower one half and the space between the tubes

Just what planned use did the Soviets have for the aerodynamically advanced *DFS 346* with its *Walter HWK* bi-fuel liquid rocket engine post war? It appeared that they wanted to use it for ultra high altitude research leading to a manned orbital rocket bomber.

The former *DFS 346* hung under the starboard wing of an interned USAF *Boeing B-29* bomber. The *B- 29* carrier flying machine would take it up to high altitude and release it...similar to how the *Bell X-1* was carried when it flew beyond the speed of sound in October 1947.

would be covered so that the required shape was obtained. For fabrication reasons, *Sänger* subdivided these tank tubes into the actual containers by means of cross-walls. The purpose of the cross-walls was to have separate containers of correct capacity one behind the other for fuel and liquid Oxygen. Each fuel tank would lie symmetrically with respect to the axis of the aircraft so that thermal stresses and twists would not develop due to external forces. *Sänger* chose a subdivision of the tube tanks that would have the effect of reducing the liquid pressure on the rear end of the tank during acceleration and to prevent the aircraft from becoming tail heavy as the tanks were emptied. Finally, *Sänger* claimed that the cross-walls would give the thin walled fuselage the stiffness necessary for taking up the dynamic flight loads at the roots of the wing spars. *Sänger* thought it advisable to put the oxygen tanks in the front end of the fuselage so that the force of acceleration driving the liquid oxygen to the pumps would be as large as possible.

Between the wings and tanks would be the bomb bay, suitably sized for projectiles up to 30-ton [60,000 pounds] in weight. When the bombs were released, the floor of the bomb bay would be opened completely for a short time so that the axis of the bomb is parallel to that of the aircraft. A dangerous situation would occur if the bomb became perpendicular, because premature detonation might occur.

Just in front of the area where the rocket bomber's main wing spar meets the fuselage, there are bays for the extendable main landing skids. These gear wells were to be partly on the vertical side fuselage wall and partly recessed into the tube tank cylinders.

Because of the rocket bomber's smooth external shape, forward visibility from the cabin probably would have been very poor. In free flight at high velocity, side view slits, and optical aids — some form of periscope — would have been sufficient, *Sänger* believed. For landing, however, a kind of detachable windshield was planned since the continued positive pressurization of the cabin and maintenance of the bullet-shape became less important on the landing approach.

The "*Sänger* Projeckt" as proposed by *Messerschmitt-Bölkow-Blohm* (MBB) in the 1960s and viewed on the tarmac as seen from its port rear side. The reusable *Sänger* personnel/cargo transporter is positioned on the back of the mother carrier/booster.

Launching and Climb

Sänger wanted to launch his rocket bomber with the minimum of fuel consumption and give it the highest velocity necessary to carry it through its long glide path. He chose the catapult-launch takeoff as the most favorable method. His rocket bomber would be brought to a takeoff speed of 500 meters/ second [1,642 feet/second or 1,075 miles/hour] by means of equipment on the ground in order to increase — as much as possible — the final velocity attainable with its on board fuel and payload. He also wanted to minimize difficulties associated with the initial wing loading of 800 kilograms/meters^2 and to eliminate the uncertain flight characteristics of the airframe in passing through the speed of sound. But equally important, perhaps moreso in fact, *Sänger* had data from medical research about the physiological responses of men experiencing high rates of acceleration. He knew that acceleration is limited only by the strength of the aircraft and by the ability of the human body to withstand it. Certainly an unmanned rocket aircraft could be driven at higher rates of acceleration than a manned machine, however, the unmanned machine also soon reached the point where greater acceleration is compensated for by the greater vehicle strength (and weight) necessary for machines capable of undergoing rapid acceleration. In addition, the rocket bomber, because of its long range and the accurate navigation necessary for the bomb release, could not be flown successfully without a pilot on board. Thus, *Sänger* knew that acceptable rates of acceleration would be limited to values that could be withstood by trained pilots.

The most recent studies available to *Sänger* in 1944 indicated the highest accelerations bearable by a man in different body positions was a function of the length of acceleration. It had been found by *Dr.-Med. Strughold* that for a pilot in the normal sitting position, this limit was determined by disturbance of the circulatory system caused by the loss of blood from the brain, or heart, due to differences in hydrostatic level inside the circulatory system itself. This danger deceased in the prone position, where the limit then seemed to be set by difficulties in breathing as a result of greatly increased

Eugen Sänger's wood garage/storage shed at the University of Vienna. Seen about 1934. _Sänger_ called this shack-like structure his "Experimental Rocket Station-Vienna, Austria."

Opposite: _Horst Schneider_ as he appeared working at Kuibyshev, USSR turbojet development engine works 1945 into the early 1950s.

Former *BMW* turbojet engine specialist *Horst Schneider,* who was also a guest of the USSR 1946-1954, told this author in May 1989 that *Eugen Sänger's* sub-orbital rocket bomber had little support in *Nazi* Germany during *World War II*. This was, said *Schneider*, because it was just too far out on the horizon of possibility. *Sänger's* efforts, however, were mostly paperwork with some fundamental experimentation related to expected problems with body heating during reentry and fiction heat on the skids during landing. *Schneider* told this author that he once saw a project drawing for an orbital bomber in the USSR, and it appeared that *Sänger's* vehicle provided for the installation of two of *Wernher von Braun's A-4* (*V2*) liquid rocket engines. The Soviets also considered the use of solid propellant rockets. *Eugen Sänger's* wife, *Irene Bredt-Sänger*, performed research work with metal dispersions for solid propellant rockets to increase their specific impulse. *Schneider* told this author that he also viewed in Russia former German *GEKADOS* (*geheime Kommandosache*) reports on the *Sänger* Project translated into Russian. These translations had been issued by *ZIAM* and *ZAGI*. To *Schneider's* knowledge, the Russians never picked up on *Bredt-Sänger's* work. However, they offered to make any German a very rich man if they knew where *Sänger* and his wife could be found postwar. The *Sängers* were probably aware that there was a price on their heads, so they left Germany for France immediately after the surrender. In France they were welcomed in the aerospace industry. Also the *General Electric Company*, here in the United States, picked up on *Sänger's* work with a similar project to serve as a reusable space booster in the 1950s. *Schneider* said that he also saw a report relating to *Sänger's* project in Huntsville, Alabama, in1958. The author of these reports was a German by the name of *Kappus*, who had emigrated to the United States shortly after the war with *Dr.-Ing Anselm Franz*, the former head of the *Junkers Jumo 004* turbojet engine. *Franz* was for many years head of gas turbine development at *Lycoming* in Connecticut. He knew *Eugen Sänger* personally. Told to author June 23rd 1989, Covina, California. Photo by author.

Eugen Sänger (right) and assistant *Stefan Sztatecsny* (left), seen filling a liquid oxygen container at their rocket test site in Vienna, Austria, in 1934. Notice the complete absence of protective clothing, not even gloves or eye protection!

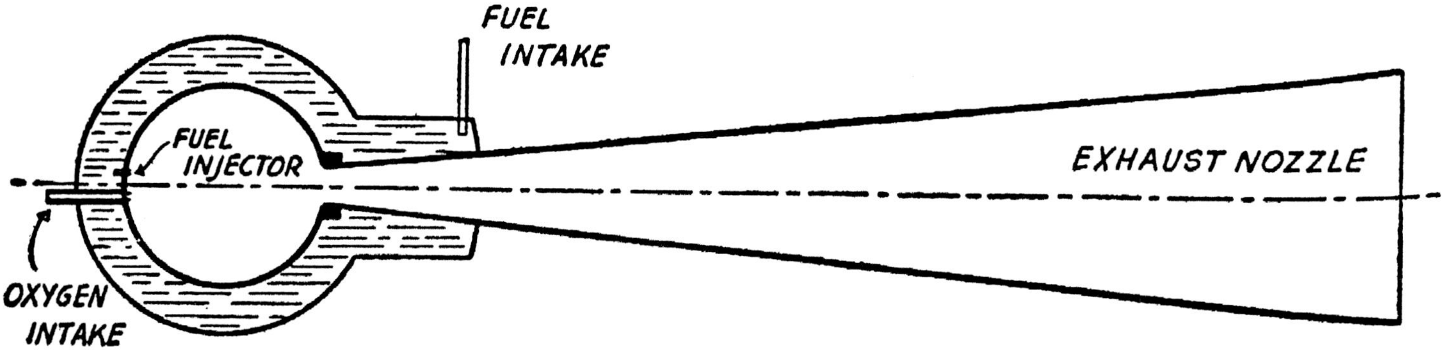

A pen and ink illustration of the prototype scale *Sänger* 100-ton thrust rocket motor. This prototype's combustion chamber had evolved into a sphere with about a 2 inch diameter. Its exhaust nozzle was about 10 inches long, with the muzzle opening equal to the diameter of the combustion chamber. The combustion chamber was cooled through the use of a cooling jacket with fuel circulating under pressure. Liquid oxygen (lox) was fed into the combustion chamber at high pressure. Notice the filter injector just above the lox injector.

weight of the thorax. *Sänger* believed that the most favorable flying position during hard acceleration was the prone position with the head facing forward. Research at that time indicated that accelerations due to the rocket engine would be best absorbed by the pilot reclining in a vehicle in a direction perpendicular to the axis of his body. In the prone position in *Dr.-Med. Strughold's* centrifuge tests, acceleration of 20g had been withstood for more than a minute by drugged primates and accelerations of 17g for more than three minutes by male volunteers.

During the rocket bomber's takeoff and climb, there would be two phases during which the vehicle's acceleration and 'deceleration' could reach critical levels: The catapult takeoff and then toward the end of the climbing path. The catapult process at takeoff, *Sänger* estimated, would last several seconds. It would start rather suddenly, peaking quickly to a maximum rate for about 11 seconds, and then would stop suddenly as fuel burnout occurred. It appeared to *Sänger* that adjustment of the pilot's circulatory system would be scarcely possible. Because of this impulse and rather long ac-

***Irene Bredt* in the mid 1940s, when she was one of *Sänger's* two research assistants at Trauen/Fassberg. *Count von Helmut Zborowski* was *Sänger's* other research assistant.**

***Eugen Sänger* in the early1940s.**

Count von Helmut Zborowski **in the mid 1940s.**

tion, he believed a pilot could not go beyond acceleration at takeoff of more than 5g, even for the favorable perpendicular sitting position. Then again, higher takeoff accelerations, though they result in economizing on starting-rockets by decreasing the run on the ground, were considered to be unfavorable as regards flight performance, since they necessitated stronger construction of the fuel tank assemblies.

Conditions were quite different at the end of the climb. During the climb, the acceleration increased in the ratio of Go/G – the result of decreasing mass of the aircraft compared with the constant motor thrust. The thrust of the rocket is constant, but at the start of the rocket burn, the mass is greatest, so acceleration is lowest. At the end of the rocket burn, the mass is low, so acceleration is high. Even if the aircraft flew without payload, this ratio would end up at

An overhead view of the *LFA's "Hermann Göring Institute"* at Trauen-Fassberg in the early 1940s.

10. This climbing process would have lasted several minutes, with the greatest acceleration occurring in the final seconds of powered flight. The human body has plenty of time to adjust itself to high acceleration and the highest acceleration expected here would last for only a short time. These human conditions, *Sänger* believed, were very favorable and the effect of the high acceleration on flight performance also was favorable. *Sänger* came to the conclusion that a pilot could withstand the shock of the sudden, initial accel-

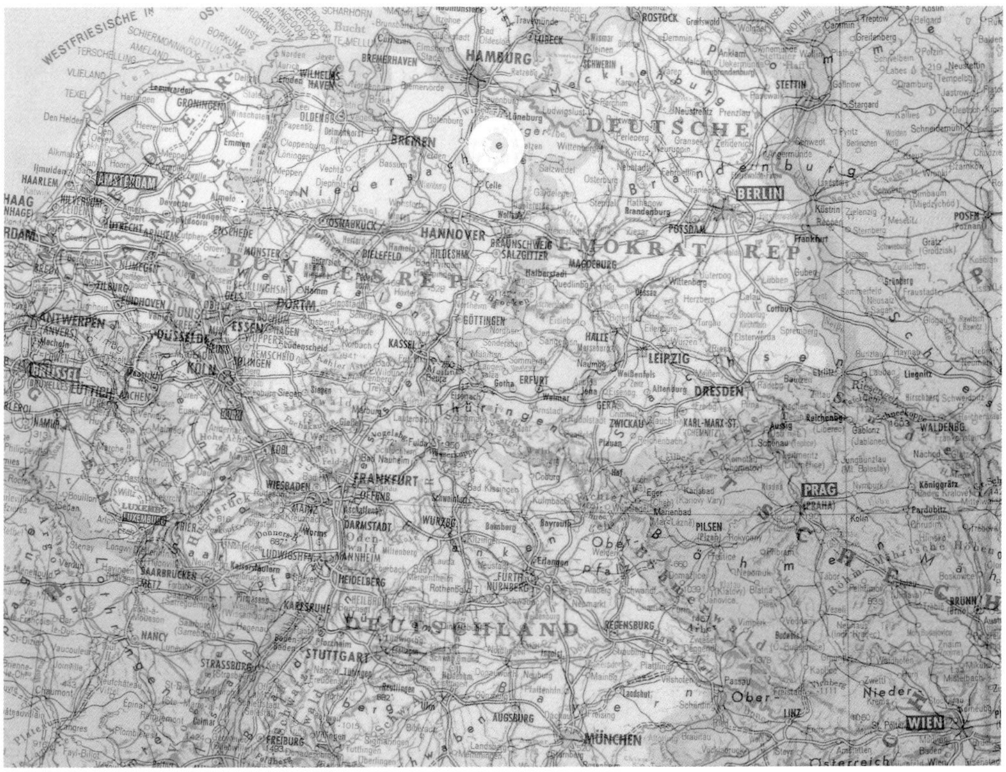

A general map of Germany. The "circle" seen in the upper top center of the photograph represents the approximate location of Trauen-Fassberg within the Lüneburg Heide where *LFA's "Hermann Göring Institute"* and *Sänger* were located between 1942 and 1944.

The way the Lüneburg Heide area appeared prior to the construction of *Eugen Sänger's LFA* rocket test facilities at Trauen/Fassberg.

eration and in view of the perpendicular seating position of the pilot, allow a design acceleration rate of 10g for his rocket bomber.

Sänger also calculated that the limiting acceleration rate of 10g might not be reached if the bomber varied its cargo load. For example, if it carried only 5-ton [10,000 pounds] of bombs on board, then its acceleration rate limit would be 6.67g. The bomber's acceleration at the start of the climb was expected to be 1g and increase after 340 to 360 seconds to 6.67g or even 10g, if – as *Sänger* said – frictional forces proved negligible. Nevertheless, the last increase from 6.67g to 10g occurred in the very short time of 20 seconds. Only during these last 20 seconds, *Sänger* knew, would the pilot be subjected to a potentially critical rate of acceleration, after already becoming accustomed to the high acceleration gradually and uniformly experienced during the previous 340 to 360 seconds. Fur-

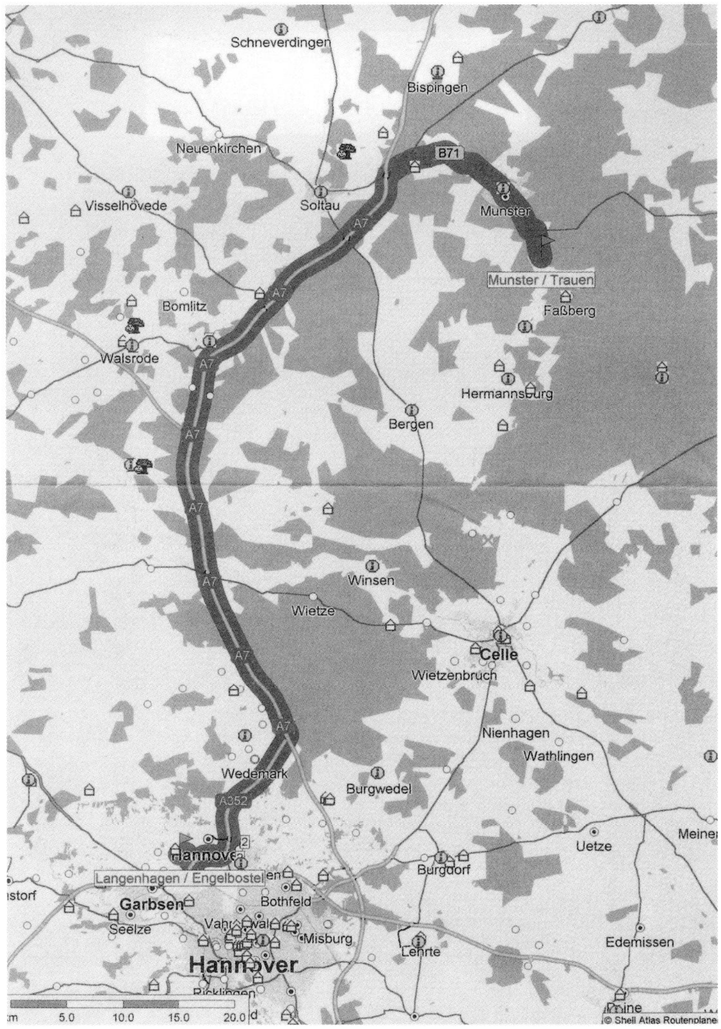

Trauen/Fassberg is located about one hour north of Hannover on Autobahn #7 as shown on this map of the area.

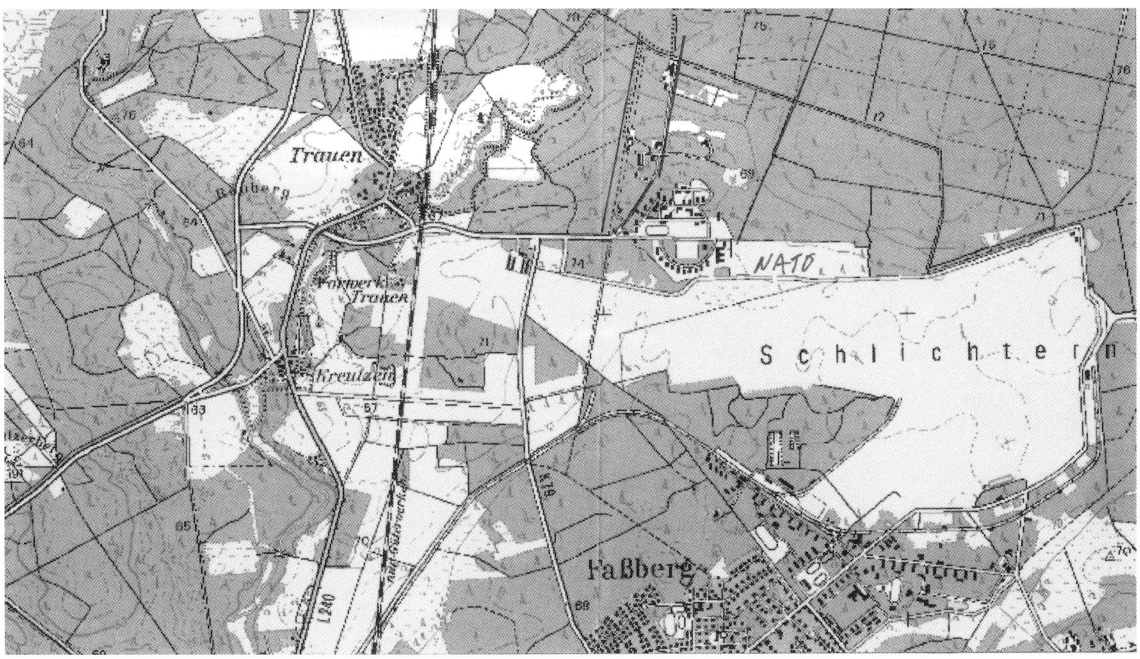

A site map featuring Trauen and its sister village of Fassberg located to the east and south.

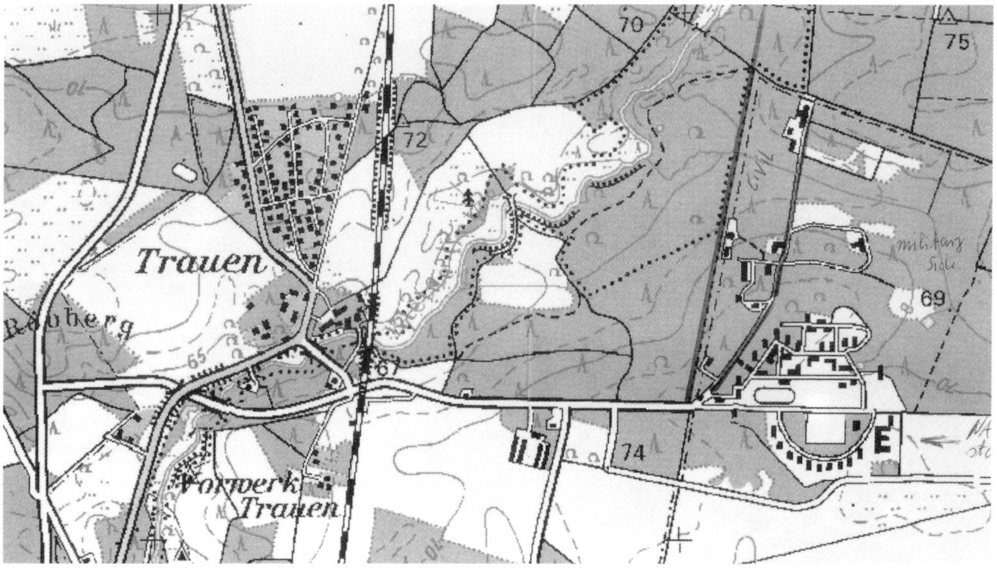

A site map of the small village of Trauen as of the year 2001.

thermore, *Sänger* believed that a possibly intolerable physiological risk to the pilot would only occur in the rare case where he took off without a bomb load. *Sänger* felt confident that the acceleration rates a pilot would experience during takeoff and climb in his rocket bomber carrying a normal weapons load would be within the limits of human tolerance.

Catapult-Launching The Rocket Bomber

Sänger's rocket bomber would have to be brought to a takeoff speed of 1,075 miles/hour [1,642 feet/second or 500 meters/second]. For this purpose, a catapult-like, perfectly straight, horizontal launching-track several kilometers in length would have been used. On this track would have sat a sled which would support the rocket bomber. The sled would be propelled by a rocket apparatus having a large thrust but moderate exhaust speed. *Sänger* estimated the rocket bomber, its takeoff sled, and the launching rocket/fuel subassembly would have weighed about 150-ton [300,000 pounds]. In addition, there would be large forces acting vertically upwards due to lift and weight in the direction of the takeoff path and the thrust and drag forces, respectively. These forces, said *Sänger*, would re-

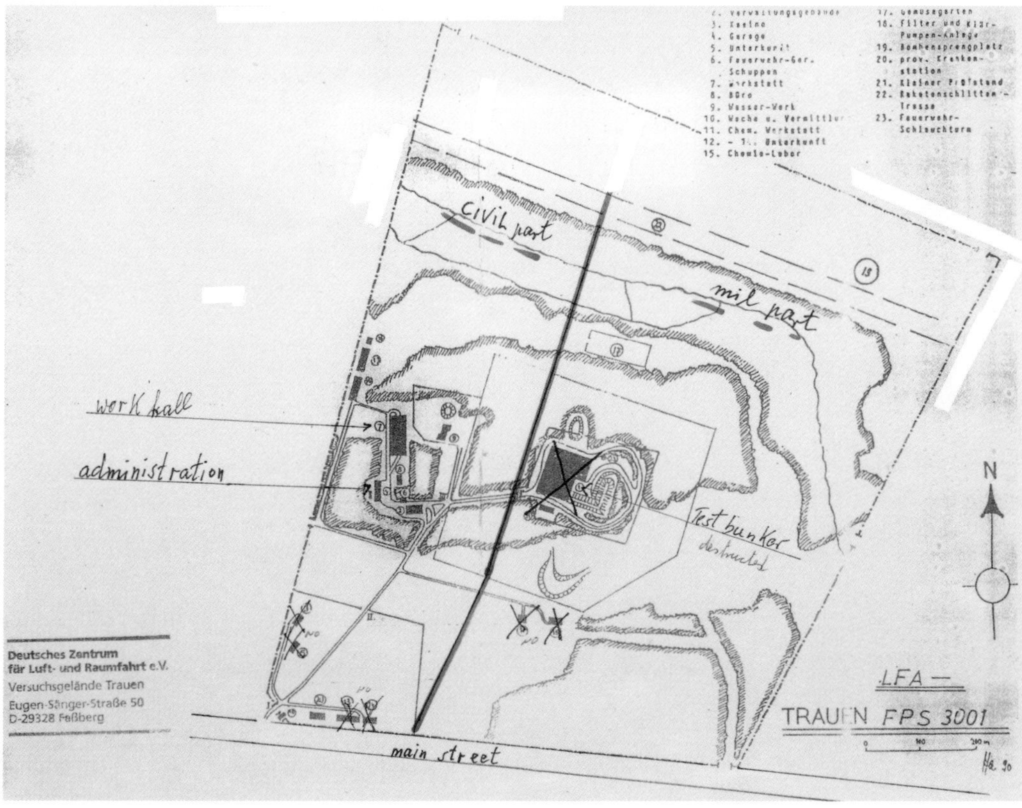

The remains of *Eugen Sänger's LFA* Trauen/Fassberg facilities as of the year 2001. However, several structures exist from the time of *Sänger*, including a large work hall (shop) and administration buildings. *Sänger's* rocket testing bunker has been demolished by the British postwar, and all that remains are its concrete foundations.

Eugen Sänger's 2nd story offices at Trauen/Fassberg exist, and they are outlined in this photo. Photo by *Reinhard Roeser.*

These are the offices for *Eugen Sänger's* former support personnel employed in the great work hall (shop) seen in the background at Trauen/Fassberg in the early 1940s, but now empty in the year 2001. Photo by *Reinhard Roeser.*

quire the precise installation of the slide rails, suggesting the single track arrangement would be fabricated out of reinforced concrete with a cross-section in the shape of an isosceles triangle. The base of the triangle would serve to fix the structure in the ground, while the sled carrying the aircraft rode on the apex of the triangle. The construction of the launch track arrangement would consist of a large main rail lying on top of the wall. This rail would be configured to restrain the enormous lift forces of the accelerating aircraft pulling up against it, but also to provide limited restraint against the massive force of drag coming against the combined aircraft and booster rocket assembly as it moved forward. In addition, there would be two lead-rails halfway up the wall which would support only occasional small forces and which would stabilize rotation of the aircraft and launch assembly around its lateral (roll) and longitudinal (pitch) axes. Low viscosity lubricants would be used on the main and lead-rails so that the sliding faces of the sled could glide on them with little friction. *Sänger* said that the sled itself was to have been mounted on the sliding faces attached to each wall and would have had a pair of fittings to prevent overturning. The aircraft to be catapulted would sit up front on the main body of the

These are the offices for *Eugen Sänger's* former supporting engineers remaining at Trauen/Fassberg in the year 2001, but are now empty. Photo by *Reinhard Roeser.*

Eugen Sänger's great work hall (shop) at Trauen/Fassberg as it appears in 2001, and empty. In this work hall, *Sänger's* technicians fabricated his experimental rocket engines for testing. Photo by *Reinhard Roeser.*

This empty building at Trauen/Fassberg in 2001 once housed emergency equipment, such as fire trucks. Photo by *Reinhard Roeser*.

These are the remains of *Eugen Sänger's* experimental rocket engine testing building at Trauen/Fassberg as of 2001. Photo by *Reinhard Roeser*.

sled. The sled's assembly of *V2* rocket engines from *Wernher von Braun* would be at the back, behind the aircraft. The high takeoff thrust would be transmitted to the rocket bomber through the main body of the sled, which would be latched to the fuselage with explosive bolts. Very high thrusts of 611-ton [1,220,000 pounds] for 11 seconds and of the utmost safety in operation would be required of the starting rockets while their exhaust speed could be moderate since they would not affect the flight performance of the rocket bomber.

Sänger figured more than 40-ton [80,000 pounds] of fuel would be required for the launch sled to push the rocket bomber along the takeoff monorail and reach a speed of 1,075 miles/hour. He did not consider self-igniting fuels such as powder, or hydrogen peroxide (H^2O^2), which were customary for moderate exhaust speeds, but he preferred a compound consisting of two materials such as liquid oxygen and a gasoline-oil-water emulsion. *Sänger* calculated the catapult-launch process required an exhaust speed of a captive bi-fuel liquid rocket at 1,500 meters/second [almost 1 mile/second or 3,600 miles/hour] so that the fuel consumption per second was approximately 4,000 kilograms/second [almost 4^2-tons/second]. He estimated that air resistence pushing against the whole catapult-launch system would be 75,000 kilograms [165,345 pounds]. On

An overall view of *Eugen Sänger's* 1-ton [2,000 pound], high-pressure combustion chamber experiment using cooling by evaporation. The propellent tanks are located above the roof to the left in the photo. The fuel pumps are directly underneath. Combustion chamber is seen in operation in the center of the photo. Notice the cloud of condensed cooling agent. The observation stand is above on the right. This rocket test stand at Trauen/Fassberg (1941-1942) could be used for testing engines up to 1-ton [2,000 pounds] thrust, and up to a firing duration of five minutes.

An overall view of a rocket motor test stand at *Eugen Sängen's LFA* research facilities near Trauen/Fassberg. These facilities were truly lavish. Seen is *Sänger's* experimental one ton-thrust bench trial of a gas-oil liquid oxygen rocket motor at 30 atmospheres combustion pressure being tested in the huge test chamber at Trauen/Fassberg up to a firing duration of 5 minutes.

the other hand, sliding friction of the sled on its takeoff rail was considered to be completely negligible when compared to the thrust and air resistance forces mentioned. *Sänger* calculated that if the required acceleration at the end of the takeoff should not exceed 50 meters/second^2, then he would need an initial push thrust of 611-tons, based on a total sled mass of 105,000-ton [210,000 pounds]. This mass included the sled, the *Sänger* rocket bomber, and the sled's launch rockets and fuel. Summarizing:

- Multiple takeoff rockets develop a combined thrust of 611-ton for 11 seconds;
- Sled takeoff mass: 105,000-ton;
- Takeoff fuel consumed: 43.7-ton;
- Length of takeoff monorail: 2,750 meters or approximately 3 kilometers (1.86 miles);
- Velocity of takeoff sled at the end of 3 kilometers: 1,500 meters/second (roughly 1,075 miles/hour);

***Eugen Sänger's* experimental 1-ton [2,000] pound thrust rocket motor test stand at Trauen/Fassberg in 1939. This rocket motor was fueled with suspended aluminum dispersed in gasoline, oil, and liquid oxygen. Aluminum oxide has formed a white fog seen here in the supersonic efflux jet.**

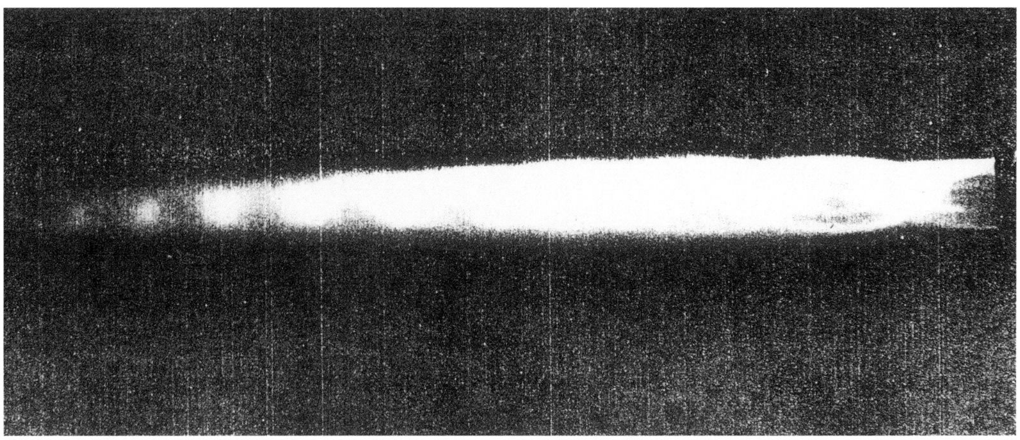

A poor quality photo of supersonic exhaust gases from the nozzle of a *Sänger* rocket motor seen during an experimental test run. Its thrust was 1-ton [2,000 pounds]. Notice the compression lines toward the end of the visible flame.

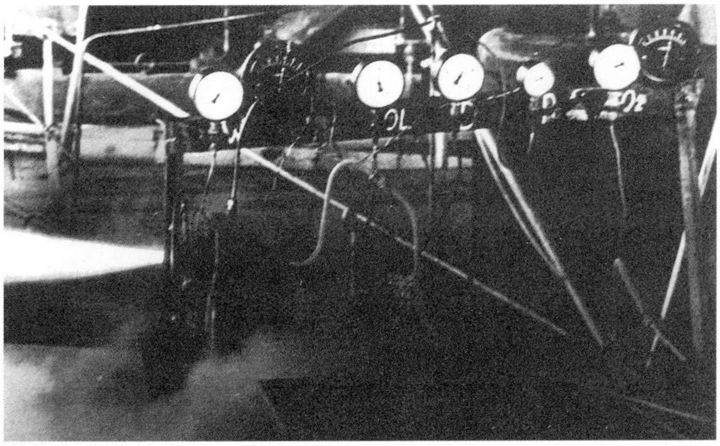

A poor quality photo of *Eugen Sänger's* bi-fuel liquid rocket engine combustion chamber tests, Trauen/Fassberg, Germany. March 1941. This particular test featured a small water cooled combustion chamber and test instruments in a duration test. Water which had been heated to 400 degrees Centigrade created 100 atmospheres pressure in its cooling system.

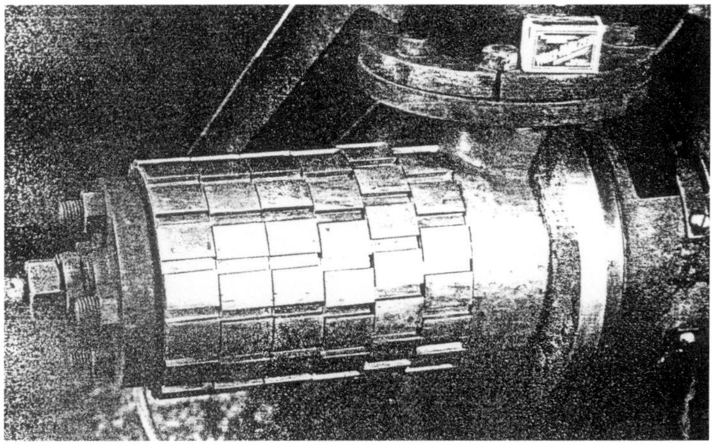

This is a photo of a miniature experimental high pressure liquid oxygen (O_2) pump designed by *Sänger* which had a 6-stage rotary pump with external bearings. At 15,000 revolutions per minute, it was capable of pumping 11 pounds [5 kilograms] per second of liquid O_2 at 150 atmospheres. Notice the match box siting atop the pump at the upper right hand corner of the photograph. *Sänger* said that this pump had proven its suitability and reliability in hundreds of experiments conducted at Trauen/Fassberg.

- Takeoff acceleration 40.2 meters/second^2 to a permissible limit of 50 meters/second^2;
- Average takeoff acceleration: 45 meters/second^2;
- Rocket bomber lifts off the launch sled at 1,500 meters/second (1,075 miles/hour);

Details Of *Sänger's* 1.86 Mile-long Monorail Launch Track

Sänger believed he would be able to successfully launch his rocket bomber from a monorail track 2,750 meters [9,022 feet] in length. This translates into about three kilometers or 1.86 miles. The rocket bomber would be placed on a sled propelled by multiple *V2* rocket engines, which would ride the monorail on so-called sliding "jaws". He offered a few details of the proposed takeoff system. First, the sliding jaws would be manufactured of bronze or hard steel. They

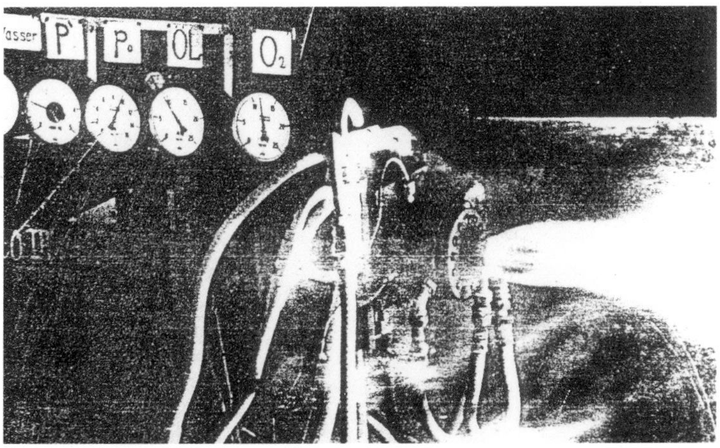

A poor quality photo featuring instruments and propellent lines. This test was run on March 20th 1941. It had a chamber pressure of 100 atmospheres, thrust of 1.1-ton [2,200] pound, and running time/duration of 3.5 minutes.

Right: A pen and ink drawing of *Eugen Sänger's* liquid oxygen/alcohol rocket engine with regenerate cooling. *Sänger* is given credit for this engineering invention, which was promptly put to use by rocketeers in other countries.

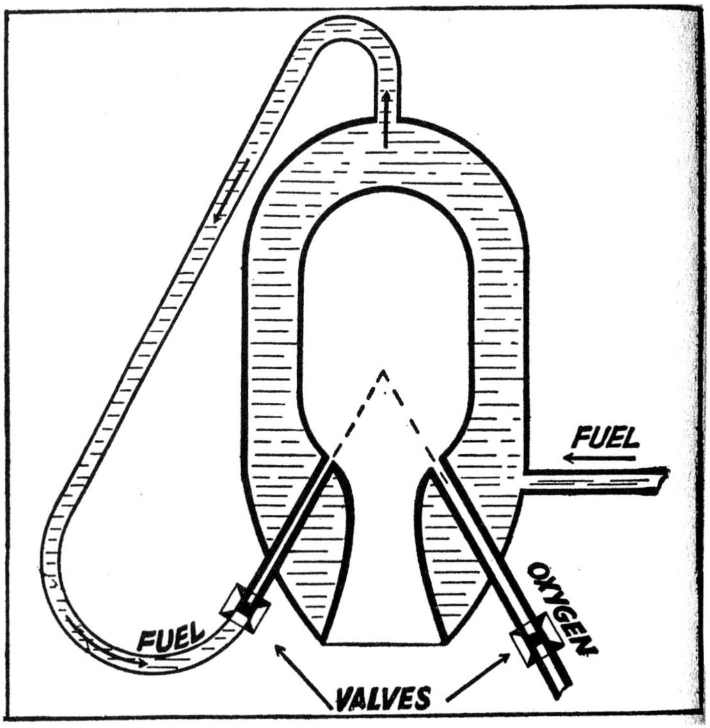

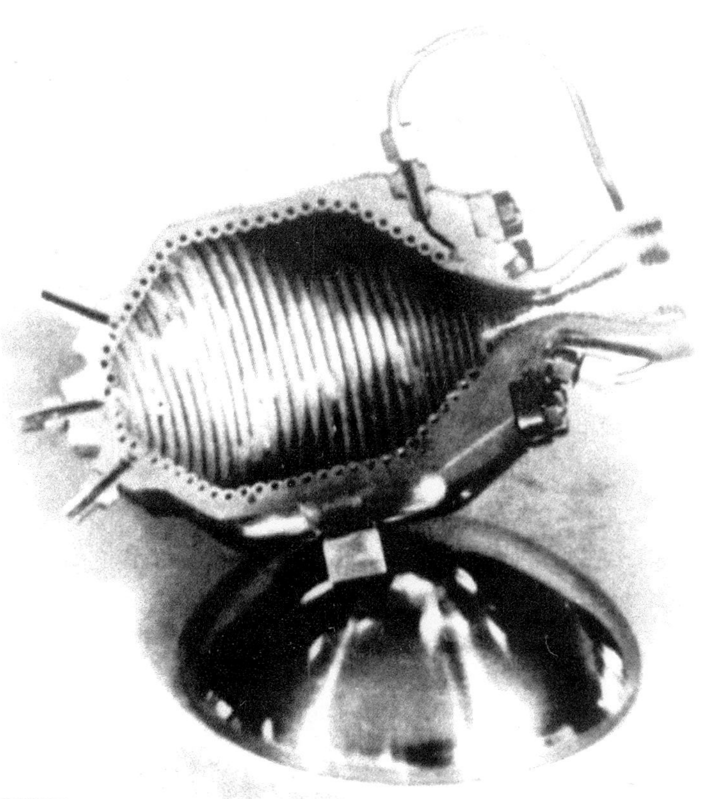

A cut away view of *Sänger's* scale model of a high pressure chemical liquid oxygen, gasoline, liquid ozone, and oil rocket engine with forced-circulation (regenerative) cooling and large aperture nozzle. This was the model for *Sänger's* 100-ton [200,000 pounds] rocket engine for his proposed sub-orbital rocket bomber.

would have a square surface of 0.25 by 0.25 meters. Their sliding surfaces would be self-adjusting and movable about a hinged edge. The minium gap between the sliding jaw surfaces and the rail surface was to be kept as small as possible to permit large loads to slide with little frictional resistance. *Sänger* estimated that the heaviest load would be applied as 'dead weight' at the beginning of the launch run — about 120 kilograms/centimeters^2 – but during take-off the load on the rail surface would drop to 4 kilograms/centimeter^2 because of the high speed and aerodynamic lift of the vehicle. In the first 30 meters [90 ^{2} feet] of the monorail, a graphite-pitch mixture would be used to support the vehicle and absorb initial frictional heat. Beyond the graphited section, *Sänger's* 3 kilometer [1.86 miles] long track's surface would be maintained and protected by a special oil lubricant carriage which would run the full length of the track along the rails. *Sänger* believed all the heat from friction generated between the rail and the sliding jaws of the sled would be absorbed by the oil. Once the sled reached the end of the monorail track and the rocket bomber had lifted off, the sled would be brought to a full stop by a caliper brake arrangement, using cast iron jaws on the sled and pressing these onto the braking surfaces of the rail. *Sänger* believed that after his rocket bomber had lifted off, the heaviest load forces on the sled would disappear and the empty sled, with its own fuel tanks now empty, could be brought to a full stop very quickly by the caliper brakes — similar to those used on railroad cars.

Sänger said if the acceleration of the rocket bomber during its launching was to remain within permissible stress limits for both aircraft and its pilot, then a stationary launch monorail of 3 kilometers [1.86 miles] in length would be required. He felt that it would not be possible to make this launching track moveable in order to

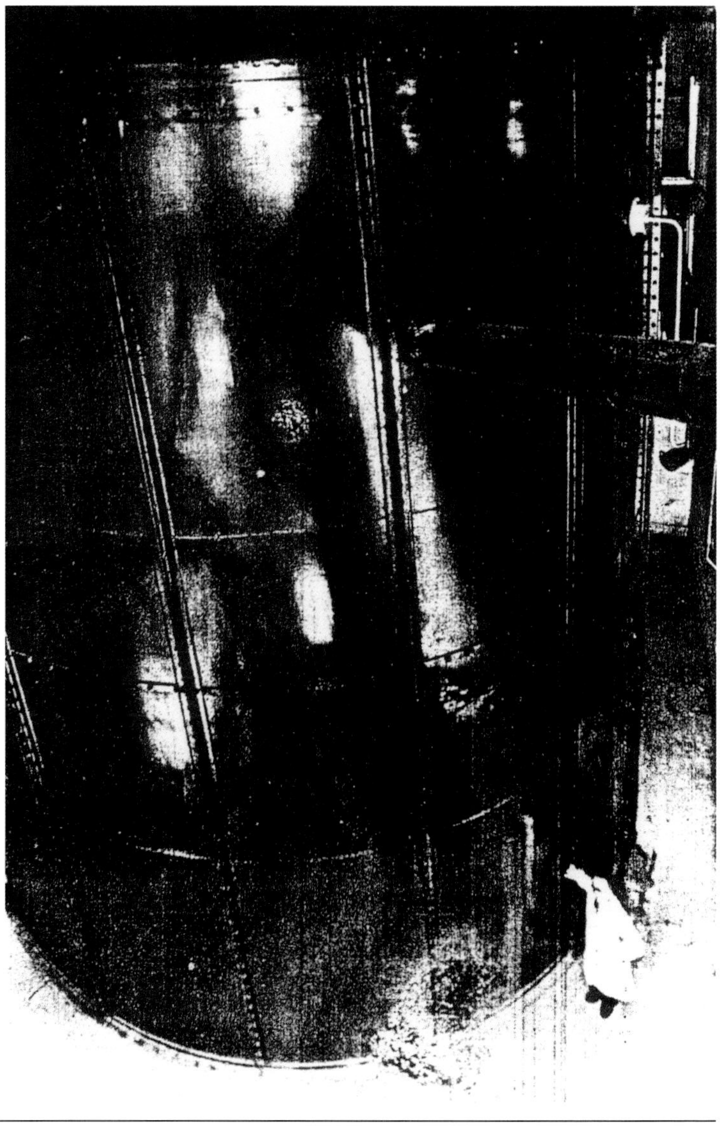

Right: Seen is a 26.2 foot high by 26.2 foot diameter [8 meters high by 8 meters] outside cover of the liquid oxygen tank at Trauen/Fassberg. A man in a white long coat can be seen in the lower right of the photograph. Inside this steel cover hung the actual container for the fluid. It was made of brass and had a capacity of 56-ton [50 meters^2] . The space between the metal shells averaged 8.5 feet [2.6 meters] in width and was loosely filled with fine powdered magnesium carbonate. The whole tank stood freely in a subterranean space at Trauen/Fassberg where it could be approached and examined from all sides. Its room had a completely normal cellar climate, and *Sänger* reported that no noticeable drop in temperature could be observed. The outer steel cover of the tank is also at normal temperature, and moisture in the air did not condense on it. The 309 pounds [140 kilograms] of O_2 vaporized each day was collected in flasks and was used for operation of the Trauen/Fassberg testing station and for welding purposes.

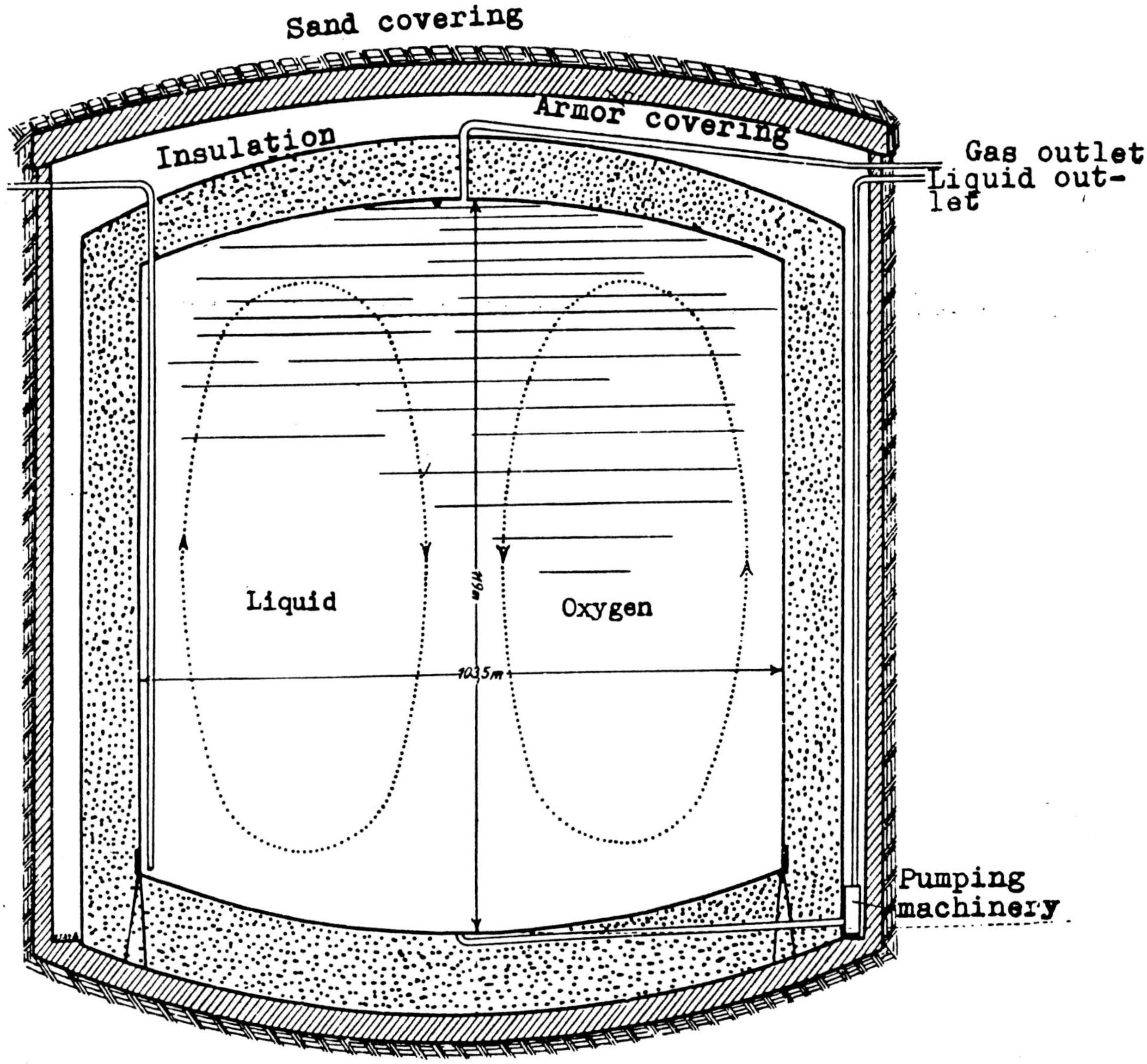

A pen and ink schematic diagram of a large container *Sänger* wanted to construct at Trauen/Fassberg. It was designed to hold one million-ton of liquid oxygen which was estimated to have a daily evaporation of 28,660 pounds [13,000 kilograms] with its 32 foot [10 meter] insulation thickness. *Sänger* argued that this container, when once filled and left alone, would become empty only after 200 years. The amount evaporated daily could be used by simply filling steel flasks and transferring them to the consumer. The cylindrical inner container of Cu-alloy would have had a diameter of 340 feet [103.5 meters] and a height of 390 feet [119 meters] which also includes its arched base. *Sänger* described the tank's bottom as conceived of as a hanging floor which would have rested on a ring running along the boundary of the outer cylinder. This ring was to have been supported on the ground by connecting pillars placed at intervals. The top cover would have been hung on the outer steel cover at various points. The insulating layer of loose magnesium carbonate powder would have been placed around the inner container, and was to have been about 32 feet [10 meters] thick on the sides and cover, about 49 feet [15 meters] thick below the floor. This insulating layer was to have been bound and supported by the steel cover which was to be rigid and strong. The whole thick-walled container would have stood on a cylinder subterranean bunker of reinforced concrete. The liquid O_2 would have come in the tank via the intake pipe to the left of the photo and then would have would have been led through the intake pipe to the floor of the container. This would have been accomplished with as little motion of the O_2 as possible during tank filling. At the top of the tank can be seen the drain pipe for the vaporized gaseous O_2 by means of which normal atmospheric pressure is maintained above the liquid surface. The liquid would have been removed at the lowest point of the container. At the same level as this point, a pump system would have stood on the right at the edge of the bunker feeding the liquid to the outlet at the top edge of the bunker.

point the rocket bomber toward the intended target at takeoff. Instead, he would install a stationary track useable in both directions so that it pointed in the most probable direction for multiple attacks, in this case east to west. More precise correction of the direction of flight could be left to the pilot during the very first part of the climb path, when the velocity and acceleration would still be sufficiently small to permit such changes of path. The rocket bomber was expected to rise up off its takeoff sled at 500 meters/second [1,075 miles/hour] climbing at an angle of 30 degrees up to a height of 4,000 meters [13,123 feet] at a velocity of 284 meters/second [932 feet/second].

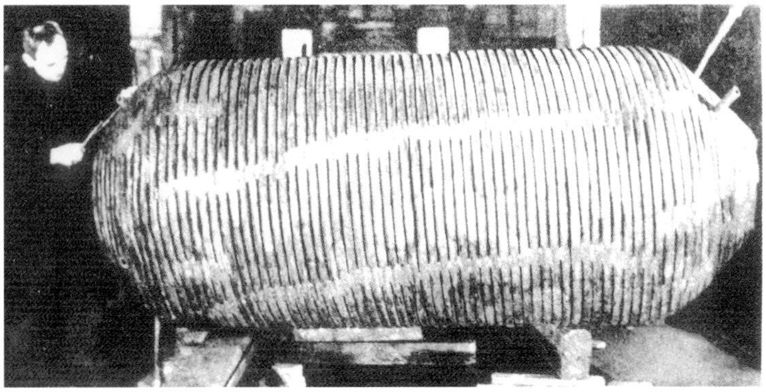

A full scale model of *Eugen Sänger's* experimental vaporization system of the regenerative combustion chamber of a 100-ton high pressure rocket for his winged, sub-orbital rocket bomber. Notice to the far right in the photo are two small diameter pipes for the injection of liquid oxygen and fuel. This model combustion chamber was not tested prior to war's end.

The Rocket Bomber's Climb Path After Catapult Launching and Lift-Off

Sänger's rocket bomber at each moment of its flight after takeoff would be directed towards a flight path profile where the "buoyant" combination of lift and centrifugal motion forces were equal to its own weight. The climb path of the rocket bomber would be determined by the forces acting on its center mass. The rotation of the Earth makes finding this trim condition more complicated. If the Earth's atmosphere were fixed in space so it didn't rotate along with the planet, then the rocket bomber's path relative to the Earth's surface could be calculated as if it were a 6th force. The atmosphere,

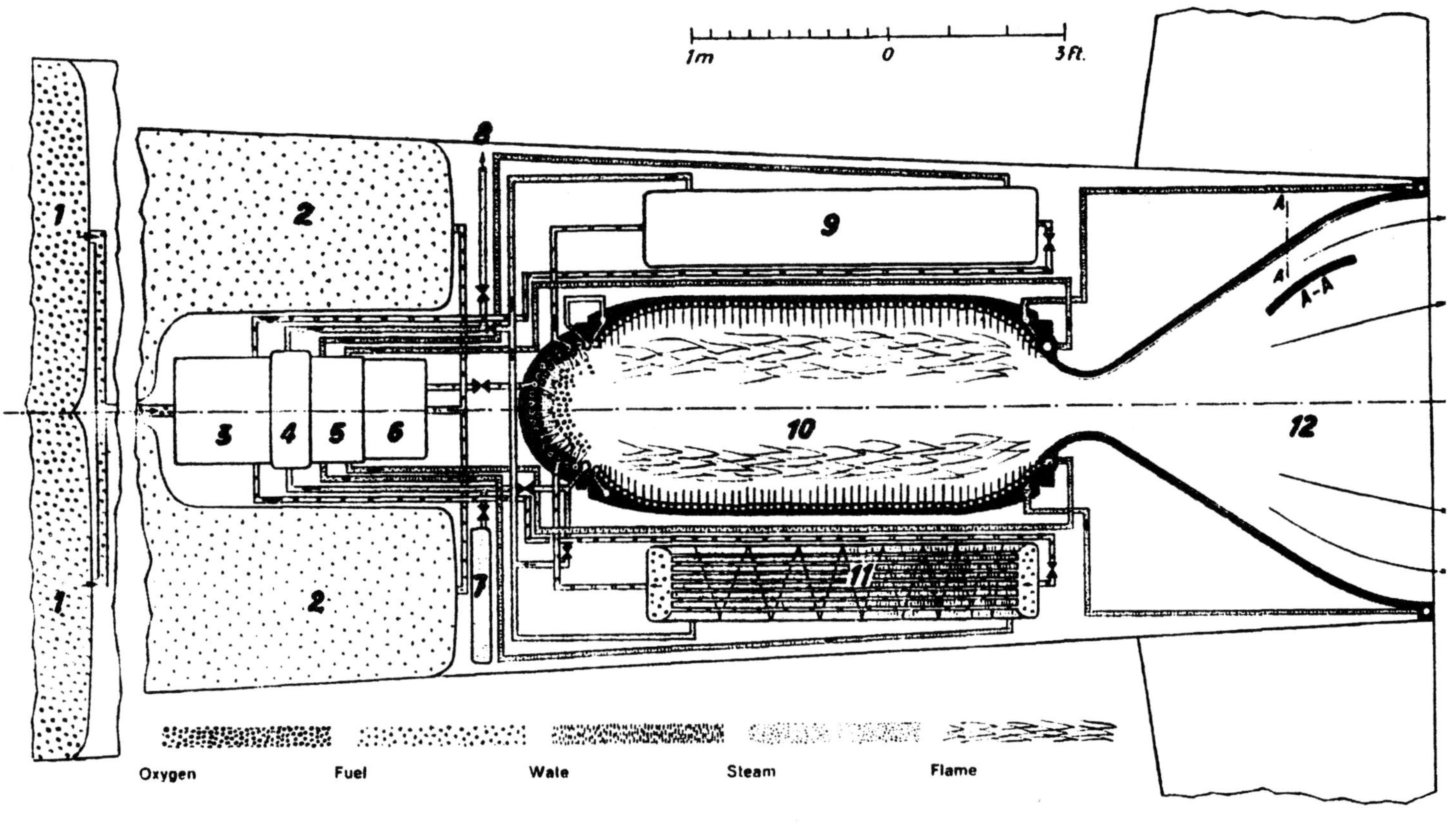

A pen and ink general arrangement of the main components of *Eugen Sänger's* 100-ton rocket engine for his proposed piloted winged, sub-orbital, reusable rocket bomber. It is drawn as seen inside the rocket bomber's rear fuselage.

01 - **oxygen tanks;**
02 - **fuel tanks;**
03 - **oxygen pump;**
04 - **tank pressurizing steam turbine pump;**
05 - **water pump;**
06 - **fuel pump;**
07 - **steam-driven starter pump;**
08 - **starter exhaust vent;**
09 - **condenser, starboard side;**
10 - **combustion chamber, simultaneously forced circulation boiler;**
11 - **condenser, port side;**
12 - **flame tube;**

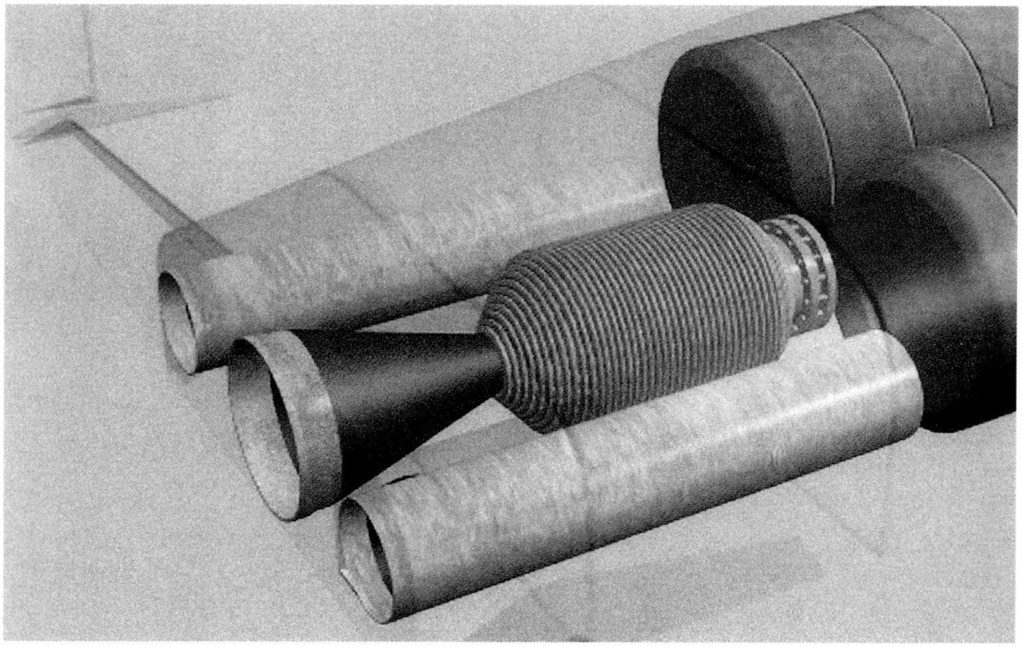

This is how *Sänger's* 100-ton thrust [200,000 pounds] combustion chamber would have appeared in his winged flying machine with its nozzle attached. A solid fuel rocket appears to the right and left of the main combustion chamber. Forward the combustion chamber are the twin fuel tanks. Digital image by *Mario Merino*.

which actually rotates with the Earth, would produce dragging forces on the rocket bomber which could cause its absolute path to be altered so that the rocket bomber might be pulled like a weather-vane by winds created by the rotation of the atmosphere. *Sänger* knew this as "the *Coriolis Force*". This interfering weather- vane action would have to be countered by the pilot's steering, if the absolute path prescribed by celestial navigation was to be maintained. The determination of the basic trajectory, *Sänger* said, was much more simple. One of *Sänger*'s associates *A. Woyczechowski*, calculated the rocket bomber's orbit:

- Begin to enter initial trajectory at 3,675 meters [12,057 feet] above sea level;
- Enter trajectory with relative velocity of flight at 284 meters [932 feet per second];
- Initial inclination of trajectory 30 degrees relative to the Earth's surface;
- Forward velocity of 4,700 meters/second [15,420 feet or 10,513 miles/hour] required to maintain attitude of "stationary equilibrium";

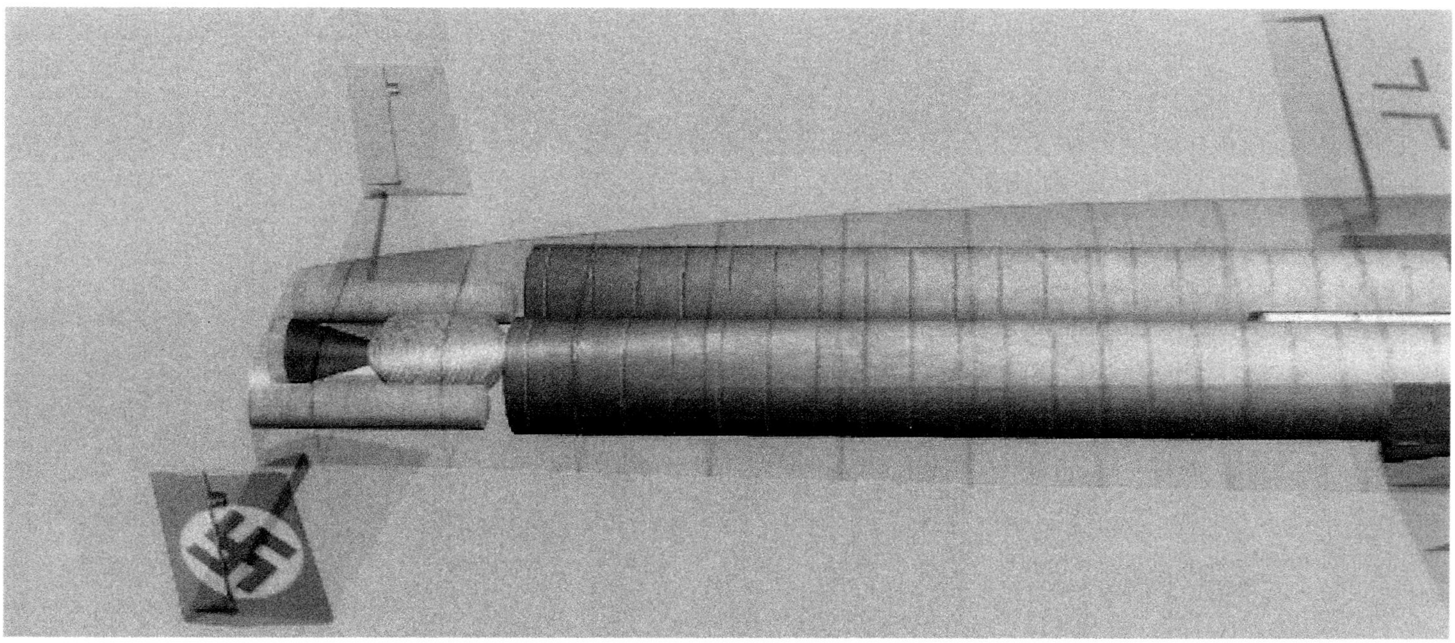

A view of the starboard aft section of *Sänger's* piloted, winged, orbital, reusable, bi-liquid rocket bomber. Between the two solid rockets is the combustion chamber. Forward the combustion chamber are the two fuel tanks. Forward the fuel tanks were the liquid oxygen tanks.

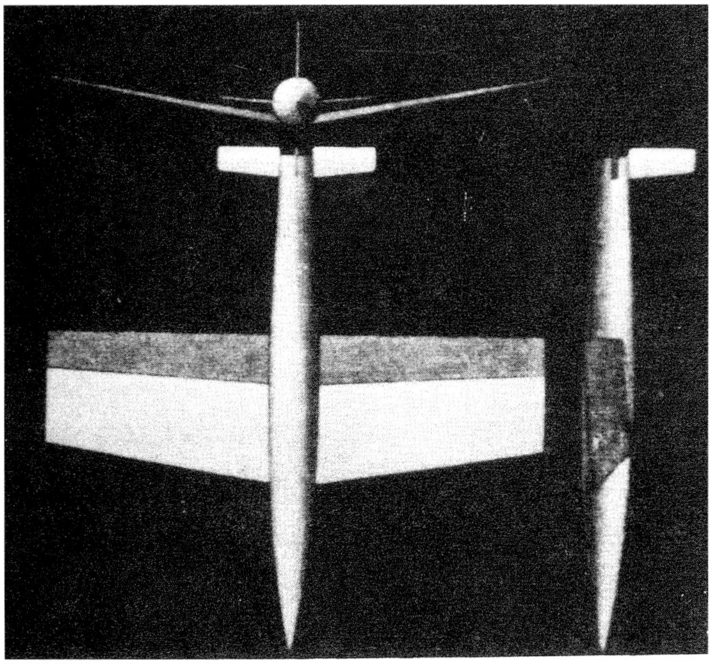

An artist's 3-view rendering of *Sänger's* proposed piloted, winged, bi-liquid rocket propelled aerospace transporter of 1931-1932.

Thus the rocket bomber would enter into a fairly low altitude, but relatively stationary condition "start window" where its "buoyant" forces were equal to its weight as acceleration continued and altitude increased. *Sänger* doubted that a true "stationary" situation would exist in the rocket bomber's flight. He described how after the bomber's launch and acceleration, its altitude at "cruise" should remain a constant distance from the Earth most of the time. At maximum altitude, with velocity still increasing, *Sänger* thought that an effort necessary to push the nose of the aircraft down to level flight relative to the Earth's surface would result in powerful centrifugal forces which could drive the aircraft above its intended stationary "equilibrium" altitude of 40 to 60 kilometers [25 to 37 miles] to an altitude perhaps over 100 kilometers [62 miles]. Any such descent from heights initially attained should be made only in a slight degree, if at all, he said, in order to maintain the forward velocity of 4,700 meters/second [15,420 feet/second or 10,513 miles/hour] which is the speed of stationary equilibrium that would allow maintenance of a 'coasting' low-Earth orbit. At this speed, *Sänger* said the rocket bomber would be in stationary equilibrium at every altitude. On the other hand, if it exceeded 4,700 meters/second forward velocity, then its pilot would suddenly find himself below the altitude of stationary equilibrium. Then the machine would have an excess of centrifugal force and would begin to climb again.

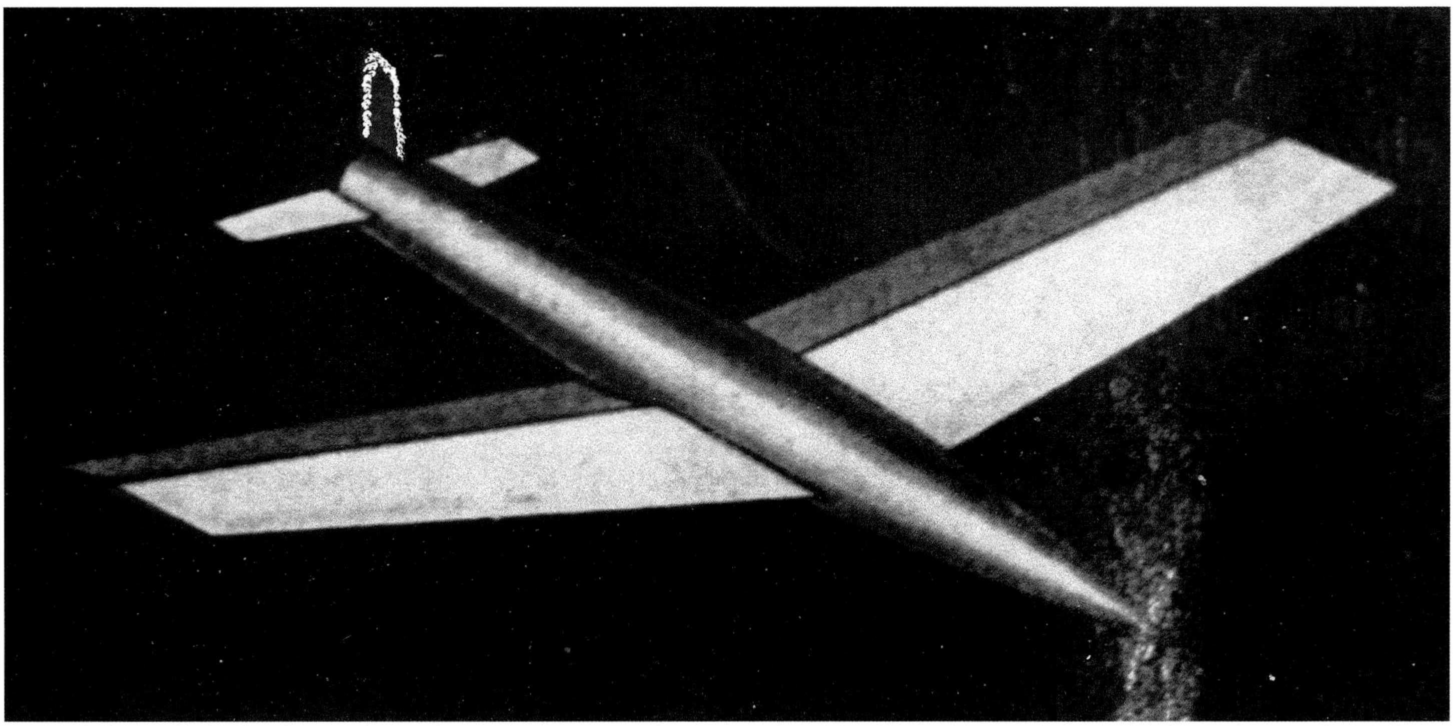

A poor quality wind tunnel scale model photo of *Eugen Sänger's* proposed winged reusable bi-liquid rocket propelled aerospace transporter from 1931-1932 as seen from its starboard underside.

An overhead planform of the proposed *Sänger* piloted, winged, bi-liquid rocket propelled aerospace transporter. Digital image by *Josha Hildwine*.

Sänger Bomber's Flight Path During Supersonic Flight

The purpose of *Sänger's* rocket bomber was to deliver bombs to any location on earth from space and then return to its takeoff site and land. Then the vehicle would be refueled, rearmed, and returned to space. To do this repeatedly, the rocket bomber pilot would have to carry out circumnavigation with the aid of precise celestial navigation. He would fly a precise low-Earth orbital path until the bomb load had been dropped. After weapon release, the pilot would either continue onward and begin to plan his descent, or the orbit could be broken off at any point, giving the pilot an infinite number of possible paths of descent and landing sites.

An important consideration for the use of the rocket bomber is how the aircraft will return to its home airbase. Unless the vicinity of the home airbase was being attacked, the bomber would simply return home by circumnavigating the Earth along its orbital gliding path after the bomb release. For nearer targets, *Sänger* said that there also was the possibility of returning by shifting course after releasing the bombs. He mentioned three possibilities:

• **First** - instead of circumnavigating, have the course home become a circle on the Earth's surface, that is, a flight along a circular cap would be more favorable than flight along a great circle, but only for vehicles below 5,600 meters/second [18,372 feet/second or 12,527 miles/hour]. *Sänger* wrote that this method of turning seemed suitable for bomb attack targets involving short ranges from the launch point;

• **Second** - the rocket bomber proceeds to the bomb target along a great circle. Once there, the rocket bomber turns hard around and returns to its home airbase along another great circle. Its turning path would take on the appearance of a spiral; a benefit would be that such tangential turns and rapid accelerations would remain constant. But the penalty of this type of maneuver would be a sharp loss of kinetic energy — 40 percent for a 90 degree turn and 63 percent for a 180 degree turn;

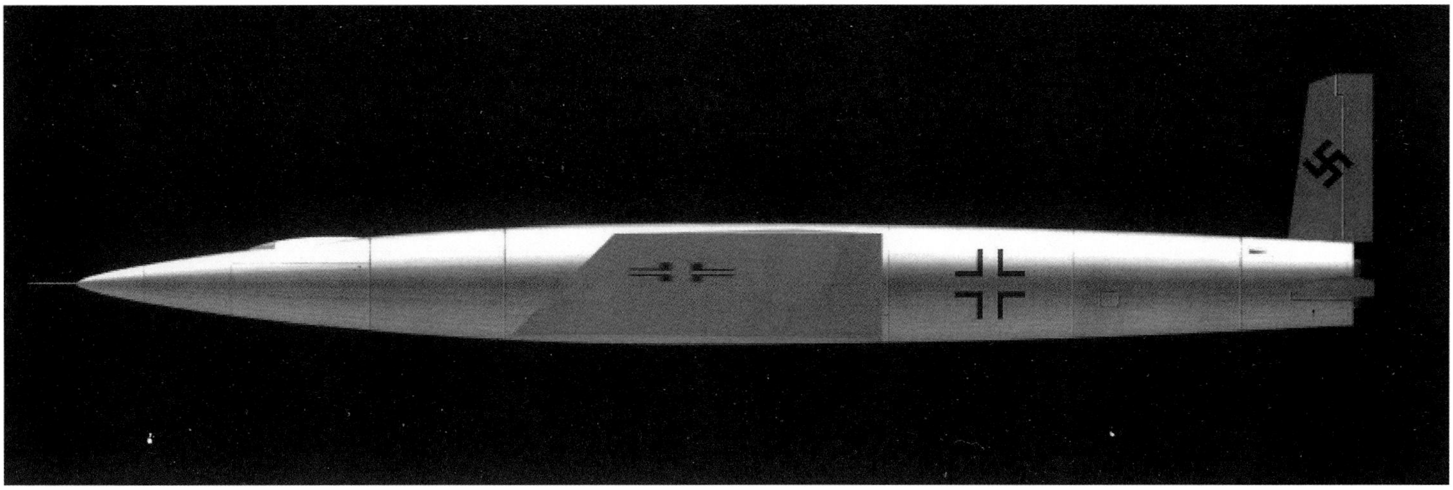

A beautiful port side view of the *Sänger* piloted, winged, bi-liquid rocket propelled aerospace transporter in flight. Digital image by *Josha Hildwine*.

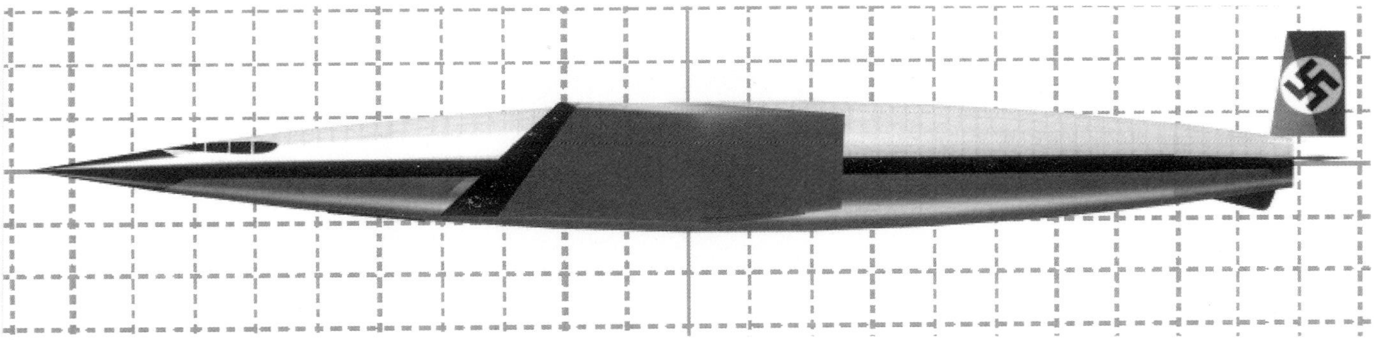

The *Sänger* winged aerospace transporter as seen from its port side. It is shown wearing the national insignia of mid-1930s *Nazi* Germany...the red band on the vertical tail and a black *Halkenkreuz* (*swastika*) within a solid white ball. Digital image by *Mario Merino*.

• **Third** - allow the rocket bomber to burn up only enough fuel during its climb to enable it to reach the bombing target. Once it reaches there and drops its bombs, the rocket bomber goes into a slow 180 degree turn and with the aid of the fuel remaining on board, the pilot reignites the motor for a second burn time, giving the aircraft enough of a push for the flight back to its home airbase. This type of flight plan — with two separate rocket burn times — has advantages. Due to the rocket bomber's relatively slow forward 'coasting' speed at altitude over the target, the pilot has the advantage of bombing with great accuracy.

The Rocket Bomber's Flight Path Returning Home and Landing

Sänger wrote that the rocket bomber's subsonic gliding flight home would begin at velocities around 300 meters/second and with the now-empty aircraft weighing about 10-ton. The machine would still have a good supply of potential and kinetic energy for its estimated 11 minute descent. The final approach to the surface of the Earth would be made at a velocity of 288 kilometers/hour [179 miles/hour] and from this point, a landing could occur. Landing aids for the pilot's forward vision would have to be employed because *Sänger* wanted his rocket bomber to touch down on a runway at no more than 150 kilometers/hour [93 miles/hour].

An overhead view of the *Sänger* piloted, winged, bi-liquid rocket propelled aerospace transporter. Digital image by *Mario Merino*.

A head on view of the *Sänger* piloted, winged aerospace transporter. Notice the severe dihedral...the inclined angle at the intersection of its wings. Its flush cockpit, as shown, is a guess, because *Eugan Sänger* gave no details. Digital image by *Mario Merino*.

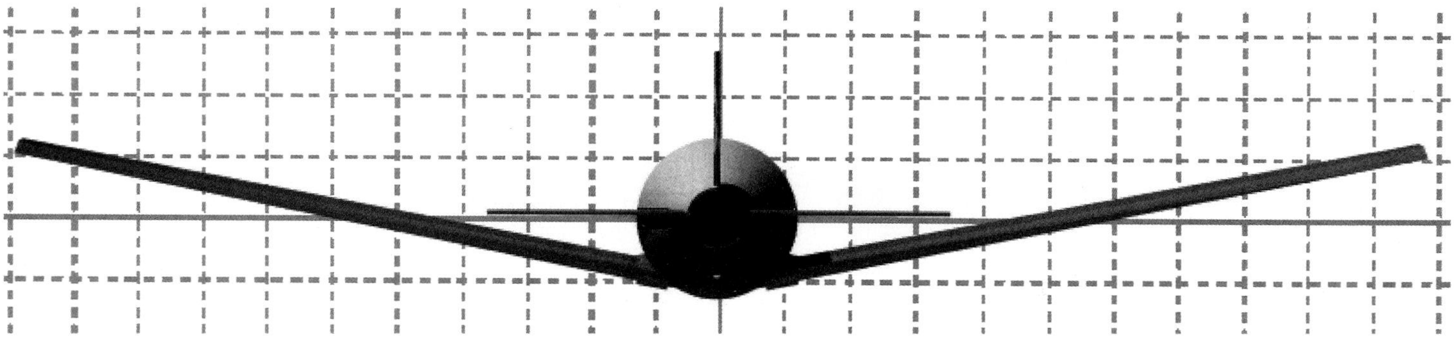

A tail on view of the *Sänger* piloted, winged aerospace transporter. Notice the small diameter of its rocket nozzle as well as the transporter's tail assembly. Digital image by *Mario Merino*.

A port side nose view of the *Sänger* piloted, winged aerospace transporter down on the concrete. Notice that this version has a metal skid attached to its nose gear. Digital image by *Mario Merino*.

The *Sänger* piloted, winged, reusable aerospace transporter as viewed from its rear starboard side. Digital image by *Mario Merino*.

Sänger designed his rocket bomber to be a tail-dragger— that is — the machine would touch down on a retractable metal skid located at rear of the fuselage. In order to slow its momentum on approach, the wings would require a high angle of attack as the machine approached the landing strip. Touchdown would occur first on the metal skid, with the nose dropping onto a two-wheeled strut as speed declined, similar to the method used by NASA's *X-15* rocket research aircraft in the 1960s. *Sänger* didn't mention the use of drag parachutes or other landing aids to slow the aircraft after touchdown. It is difficult to imagine how the pilot would have seen the landing strip below him and why *Sänger* would have chosen a skid — rather than wheels with brakes — as the means of slowing his aircraft to a full stop.

A ground level starboard side view of *Eugen Sänger's* piloted, winged useable aerospace transporter. Notice that its tricycle landing gear is fitted out with metal skids...no wheels. Digital image by *Mario Merino*.

Bombing Attacks with the Rocket Bomber

The rocket bomber proposal by *Sänger* offered to the *RLM* featured two different methods of bombing attacks: "point attacks" and "area attacks." The two procedures differed mainly in the manner and accuracy of the bomb releases.

In a "single point" attack, the bomb would be aimed precisely toward an individual target and released at low-to-moderate altitude and velocity in the same manner used by ordinary bombers. In practice, the same variety of bomb release styles were available to the rocket bomber as for other bombing aircraft — bomb release during horizontal flight, dive bomb attack, bomb release during climb, low altitude bombing, and so on. The same conditions affecting these types of release also applied to his rocket bomber, *Sänger* wrote, especially regarding the accuracy attainable and the need for adequate visibility at the target. So, *Sänger* said, nothing new could be said about this type of bombing attack.

Things were quite different for an "area" bombing attack. Here the bomb was to have been released from great altitudes...50 to 150 kilometers [31 to 93 miles] and at very high velocities of flight up to 800 meters/second [2,625 feet/second]. Since the target, for the ballistic distances involved, probably would not be visible, the release on an area would have to be aimed and timed with the aid of celestial navigation. This would allow a strike to occur completely independent of weather and visibility at the target. Because of this, *Sänger* said, "area" attacks would not have the same accuracy of a precision bomb release on a point. He cautioned readers of his proposal that bomb impact might occur several kilometers from the intended target. Using area bombing, mission planners would not intend to hit particular target points, but would plan to strike correspondingly larger areas with a sufficient probability of hitting strategic resources. In order to achieve an anticipated effect on a par-

The *Eugen Sänger* piloted, winged, reusable, bi-liquid rocket powered aerospace transporter of 1931-1932 appears very modern in this nose port side view due to its sleek lines, tapered, nose and flush cockpit cabin windows. We must remember that most government air forces in the early 1930s were still flying bi-wing machines so to these people, the *Sänger* winged rocket powered aerospace transporter must have looked like a space ship out of comic books. Digital image by *Mario Merino*.

A rear starboard side view of the *Sänger* piloted, winged bi-liquid rocket, reusable aerospace transporter. Notice its metal skids at the end of its tricycle landing gear. Digital image by *Mario Merino*.

ticular target area, *Sänger* emphasized that a single bomb drop probably would not suffice. Rather, several bombs dropped by other rocket bombers would have to be randomly distributed across the same target area so the likelihood of inflicting serious damage would increase. Still, *Sänger* believed the distribution of hits inside the target area would not be uniform, that is, the bombs would strike more frequently in the neighborhood of the desired target rather than farther away. In his proposal to the *RLM*, *Sänger* wrote that some of the bombs probably would strike hits far outside the area being attacked. However, on the basis of laws of probability, the bomb distribution could be predicted well enough so that the main goal of the attack could be achieved with the same or even greater accuracy than for a point attack. Also, the collateral psychological effects of area bombing — a type of attack for which there was no defense — would be considerable on civilian populations.

Flight Path of Bombs Released from the High-Flying *Sänger* Rocket Bomber

In order to make calculations concerning the bombs released from the rocket aircraft, *Sänger* also had to make some assumptions about the external shape of these projectiles. Best suited to the anticipated conditions of a high-speed drop from high altitude was a general "bullet-shape" with a flat base, tapered to a cylindrical profile at the rear, with the largest possible ratio of length to diameter.

Sänger wrote that if one assumed the explosives payload constituted 50 percent of the bomb's weight, then bomb a weighing 10-ton would have a length of 11.20 meters [36 3/4 feet] (or if it were 5-ton then a length of 6.16 meters [20 1/8 feet] and if 1 ton then a length of 3.60 meters [11 3/4 feet]). He said that in real life any bomb load could be made up out of one large, or a combination of the smaller bomb sizes. The considerable space required for the

A photo of *Sänger* and his assistant *R. Schmid* at Trauen/Fassberg (both seen in the far left of the photo) setting up a dynamical friction test track in May/June 1939. This was to test the basic feasibility of sliding at high speeds. He made the tests with projectiles sliding along curved metal walls lubricated with vaseline. *Schmid* had constructed a large diameter test bed consisting of a military carbine and a closed circular sliding track which consisted of a drawn-steel tube of 32.8 feet [10 meters] inner diameter which had been slit along its axis to produce a U-shaped groove. A stainless steel bullet was fired with a velocity of about 2,624 feet/second [800 meters per second] from a standard-issue *Wehrmacht* military carbine rifle into the spirally curved entrance of a circular, closed, and lubricated steel groove with U-shaped cross section. During this experiment the bullet would pass through all velocity-ranges down to complete rest. *Sänger* discovered that sliding velocities up to 1,640 feet/second [500 meters/second] were possible on a carefully finished and lubricated rail. Just how his proposed 100-ton weight orbital bomber would do on a monorail at speeds up to 1,113 miles per hour on a 1.8 mile long single steel monorail was the ultimate question.

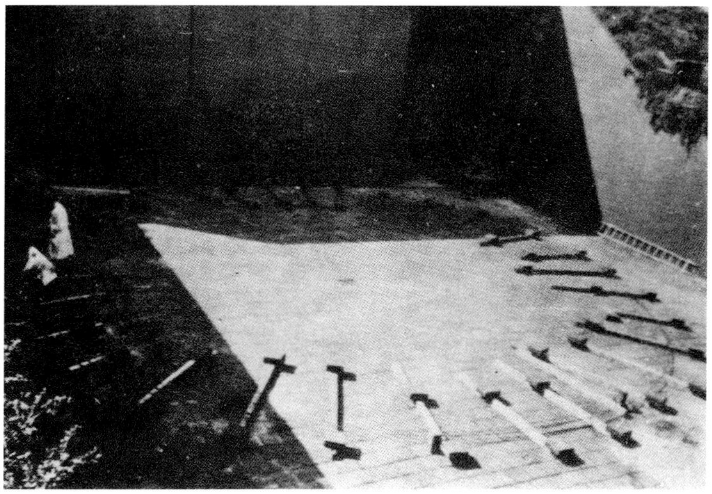

An interior view of a high-temperature 4 foot 9 inch [1.5 meter] diameter *Sänger* experimental liquid ramjet engine as seen from its rear looking straight through to its front opening. This particular ramjet is mounted on the open bed of a 2 1/2-ton *Opel "Blitz"* truck.

A starboard side view of one of *Eugen Sänger*'s ramjet experiments mounted on the truck box of an *Opel* truck. The upright gas cylinder probably contained propane gas. This gas was used as fuel for *Sänger's* numerous ramjet designs/tests.

projectiles would require an adjustment of the internal bomb bay and fuel tank-spaces, that is, a larger fuel load to achieve greater range would require decreased bomb loads and vice versa. *Sänger* believed that his rocket bomber would be more effective for point bombing attacks rather than area bombing attacks.

In estimating the path of flight of a bomb, *Sänger* said that analysis could proceed in much the same way as determining the descent path of the aircraft itself. The picture of forces acting on the bomb during launch and powered ascent is the same as that for the climbing aircraft, since it is contained within the weapons bay. After release and without any external guidance from the launching aircraft, however, the bomb could no longer be kept in a definite orbit-plane, or otherwise controlled by the pilot. So the bomb might be expected to "weathervane" while passing through various air masses during descent and also to be affected by gravitational forces caused by the Earth's rotation (*Coriolis Force*). *Sänger* wrote that these forces did not act at the center of mass of the projectile but rather behind it, particularly if the bomb had stabilizing fins. In the case of keeping the aircraft on course, these forces could easily

***Sänger's* experimental ramjets had a huge diameter, as seen in this photo. It is a 2 1/2-ton *Opel "Blitz"* truck carrying one of *Sänger's* ramjet engines. If you look closely, one can see that this ramjet is fired up. Notice the smoke pouring out the rear of the ramjet.**

Wow! *Sänger's* pioneering ramjet engine working! His experiments at Trauen/Fassberg, shown here in 1941 on the back of a 2 1/2-ton *Opel "Blitz"* truck, were low tech. However, by the time of *Nazi* Germany's surrender, *Eugen Sänger* had acquired more practical experience than anyone else in the field.

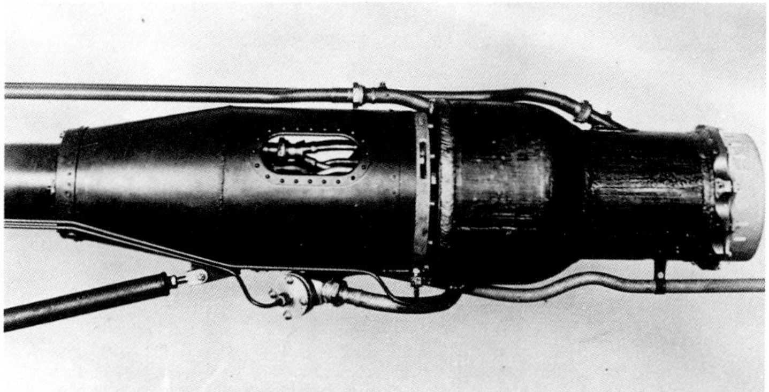

The combustion chamber of a *Walter HWK 509* bi-fuel liquid rocket engine, shown here, did not require any outside source of oxygen as did the ramjet engine. The *HWK* combustion chamber was provided with everything it needed for combustion, and thus could operate anywhere...on the ground or in the air...not like the ramjet or pulse jet engines.

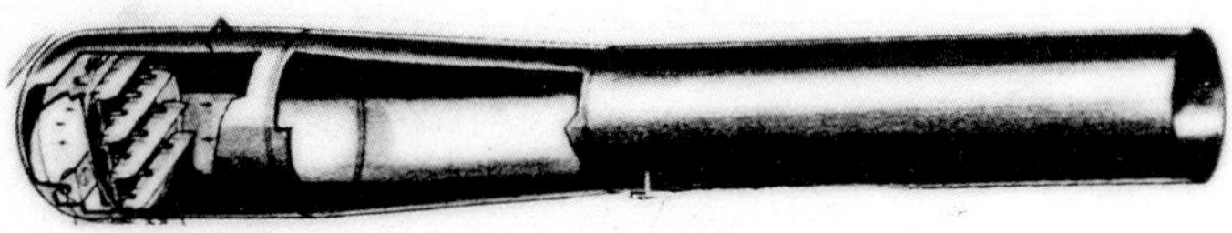

A pen and ink cut away illustration of the *Schmitt-Argus As 014* pulse jet engine. A pulse jet is similar to a ramjet, except that a series of spring-loaded shutter-type valves is located ahead of the combustion section. In a pulse jet, combustion is intermittent or pulsing rather than continuous.

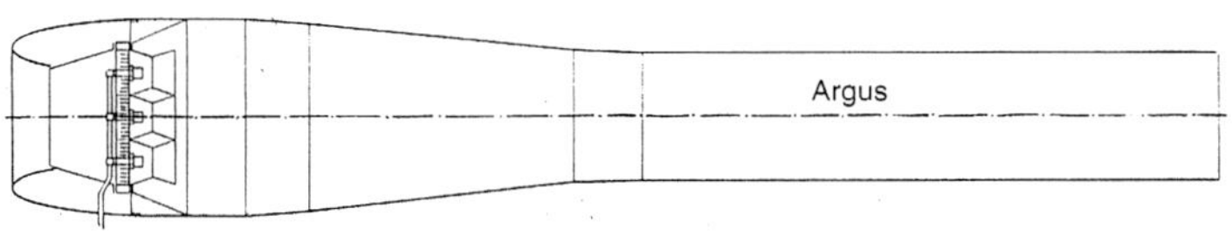

In the *Schmitt-Argus As 014* pulse jet engine, air was admitted through the valves and combustion was initiated, which increased the pressure, closing the valves to prevent backflow through the inlet. The hot gases were expelled through the rear nozzle, producing thrust and lowering the pressure to the point that the valves again opened and admitted fresh air. The cycle was repeated. The *Schmitt-Argus As 014* fired at a rate of about 40 cycles per second.

***Eugen Sänger* was an advocate of the "*Einlaug*" type of diffuser, which has all the pressure rise inside the unit as shown in this pen and ink illustration of *René Lorin's* ramjet patent of 1908. The principle of this engine mainly consists of designing a duct in such a way that air continuously streaming in against flight direction is first decelerated and compressed, then heated, and finally ejected at accelerated speed, whereby the entire procedure is completely continuous and requires no rotating parts.**

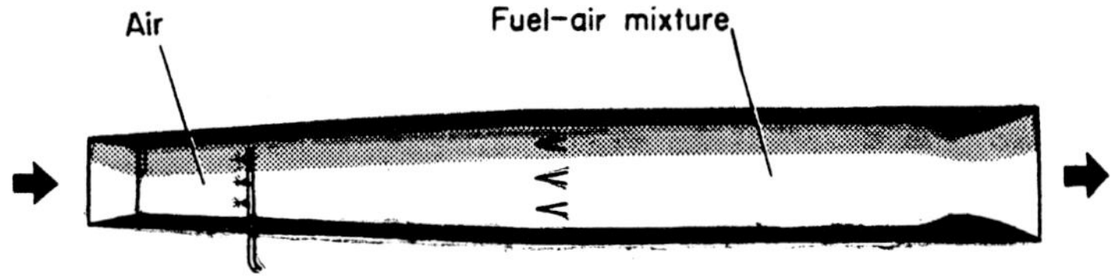

***Eugen Sänger's* ramjet engine experiments were conducted using gasoline as fuel, and he generally injected the fuel upstream, against the airflow, in order to obtain a reasonable efficiency of combustion.**

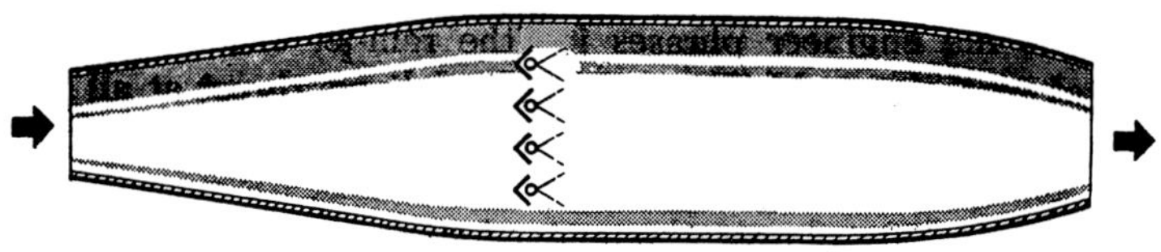

A pen and ink illustration of a supersonic ramjet engine. Notice that the burners have been located about mid-point on the ramjet engine.

be overcome by the pilot. In the case of the bomb, however, there were no such controls after release. As a result of the weather-vaning action, the bomb might veer off on a tangent course to the side. It might also start to rotate about a vertical axis through its center of mass as a consequence of the resultant torque. The flight path of the bomb would be influenced by five external forces:

- The weight and center of gravity distribution of the bomb;
- The forward air resistance encountered by the bomb in free fall after release;
- The effects of transverse weathervaning caused by descent through various moving air masses;
- The *Coriolis Force* of the Earth's rotation and gravitational field effects;
- The *d'Alembert* inertial force of the mass of the bomb itself, acting about its flight axes;

A pen and ink schematic arrangement of a supersonic ramjet engine:
1 - diffuser; **4 - fuel supply;**
2 - fuel injectors; **5 - flame holder;**
3 - ignitor; **6 - jet nozzle;**

Professor Hermann Busemann. He was in charge of all wind tunnels at *DVL*. Apparently there were some longstanding differences between *Busemann* and *Sänger* so he never had a chance of wind tunnel testing his experimental ramjet engines in the *DVL* wind tunnels as did *Helmuth Walter* of *HWK* and *Otto Pabst* of *Focke-Wulf*.

The administrative buildings of *DVL*-Berlin/Adlershof. On the right is the dining hall and lecture rooms, about 1941.

A view looking down on the open throat of one of the larger wind tunnels at *DVL*-Berlin/Adlershof. Notice the size of the two men in the lower middle of the photo compared to the wind tunnel above them.

Outside the great 155 mile/hour [250 kilometer/hour] static thrust wind tunnel at *DVL*- Berlin/Adlershof in *Nazi* Germany.

Overall view of *DVL's* facilities at Berlin/Adlershof.

A *DVL*-Berlin/Adlershof wind tunnel technician is viewing the results of an on going wind tunnel test via a thick glass port hole.

Finally, *Sänger* made some general comments concerning the accuracy of hitting a target from great altitudes at high speed. He suggested we consider the example of gunnery and the accuracy of a projectile in relation to its elevation and range of fire. For the rocket bomber, the "range of fire" would be the length of the path of fall of the bomb. For example, with a mean range of 600 kilometers drop distance to the target [373 miles], the probable 50 percent scatter would be 6 kilometers [3 3/4 miles] around the intended target impact point, that is if one could release a free falling projectile from the bomber with the same accuracy as one can fire a shell from a cannon. The actual error, *Sänger* suggested, would be the combination of errors in release which are determined by position, velocity, and direction of the bomb at release, similar errors in the bomber's attitude at the point of release, and of deviations of the bomb's flight path during descent which are caused mainly by fluctuations in the density and flow of surrounding air masses and the *Coriolis* forces acting on the bomb.

The navigation of the rocket bomber to the bomb release point would be divided into three aiming procedures:

• **First Stage** - Having the catapult-launch apparatus lie generally in the direction of the target, if possible, is important. Otherwise due to the launch rail's fixed orientation, course deviations up to 90 degree after take-off might have to be made.

• **Second Stage** - The greatest correction of errors can occur right after takeoff — during the few seconds of motorless flight, or right at the start of the propulsion period when the bomber's internal engine is ignited and the direction to the target can be set by means of the compass to within a few minutes of arc. At this point, the bomber is accelerating through dense surface air where both the aircraft's normal flight control surfaces and the vanes directing rocket combustion flow can be effectively used to quickly alter course. During the post-launch glide outside the major portion of the Earth's atmosphere, the rocket bomber could not be steered quite as accurately, since the combined effects of the Earth's rotation, the thin atmosphere, and incidental influences like small asymmetric errors in steering, fluctuations of air density, and movements of air laterally across the course of the vehicle at lower altitudes, tend to affect the intended path of the bomber and bring it out of orbit.

• **Third Stage** - During the rocket bomber's glide above the troposphere — where bomb release is likely to occur, the accuracy of bomb impact depends on the ability to make minor alternations to the course of the bomber and factor in known air mass movements and local gravitational deviations beneath the aircraft. This stage would rely on the use of astronomical methods for pointing and

***Eugen Sänger's* strong competitor in the field of ramjet engine research was *Dr.-Ing. Johannes Winkler.* Although *Winkler's* field of interest was solid fuel or carbon blocks attached to the inside surfaces of ramjets, his reputation in the field of rocket science was as thoroughly grounded as *Sänger. Winkler*, supported financially by his family, had been the 1st president and founder of the German Rocket Society (*VfR*) in 1927. He was also the editor of *VfR's* magazine...the "*Rakete.*'"**

Eugen Sänger seen testing one of his experimental *Lorin*-style gasoline fueled ramjet engines on the back of a *Opel* "*Blitz*" 2 1/2-ton truck, *DFS*-Ainring 1944.

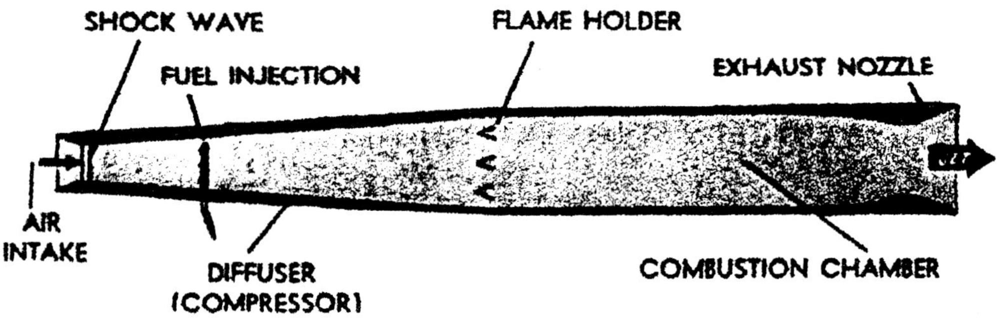

A pen and ink drawing of the port side of a *Sänger* experimental ramjet engine under development at *DFS*-Ainring. Notice its simple "stove pipe-like" design.

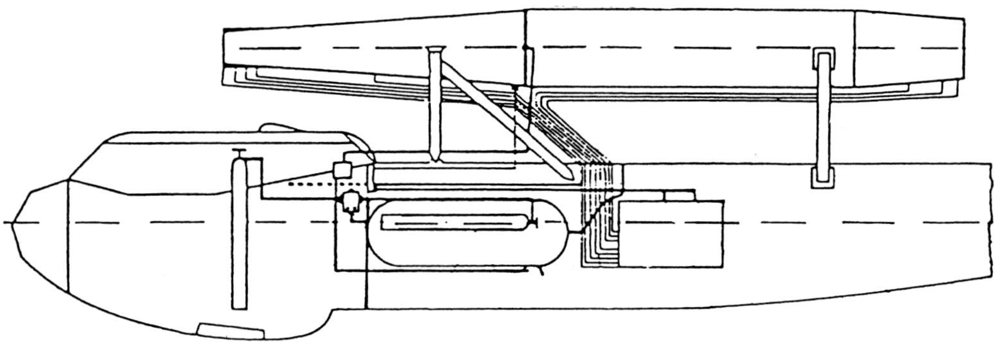

This pen and ink drawing features the fuel tank locations in the *Dornier Do 217 E-2* required for providing fuel in testing a *Sänger* experimental ramjet mounted on top of this flying machine.

Sänger was able to leave his 2 1/2-ton *Opel* "*Blitz*" truck behind when he left Trauen/Fassberg. At *DFS Professor Georgii* provided a *Dornier Do 217 E-2* for *Sänger's* personal use for tow testing his ramjet engine experiments. Notice the number "6" on the forward section of the ramjet...it is not entirely clear what this number meant. The overall length of this particular ramjet engine was 28 feet 2 inches [8.6 meters]. It had a diameter of 3 feet [1.0 meter].

As far as it is known, *Sänger* never tested his experimental ramjet engines in wind tunnels. *Pabst* and *Walter* were invited to bring their experimental ramjet units to the *Luftfahrtforschungsanstalt* (*LFA*), but it appears that *Sänger* was unwelcome. *Professor Busemann* controlled the wind tunnels at *LEA*, so *Sänger* was forced to tow his engines in flight atop a *Dornier Do 217 Z-2* aircraft owned by *DFS*-Ainring. Seen is the installation for a tow test of a high temperature ramjet engine.

would be mostly independent of the influence of the weather on the enemy target. Bomb release would be most efficient during the stationary glide because it would permit determination of position to an accuracy of a few seconds of arc, which could result in errors at the point of impact of as few as 100 meters [328 feet] from the intended target.

To determine the apparent horizon under the excellent visibility conditions available, *Sänger* suggested that the rocket bomber pilot could use a theodolite to sight on three points of the natural horizon so that vehicle altitude could be determined. When the apparent horizon and altitude had been fixed, maintenance of the prescribed orbit in space could be assured by choosing one or several stars in a line perpendicular to the horizon through the position of the target, then following the apparent motion of one or more stars with the theodolite.

If the guiding star remained in the plane of orbit on the vertical cross-hair of the theodolite, the pilot would know that he was in the prescribed orbit-plane. On the stationary flight plan, the pilot could determine his absolute velocity from the apparent motion of the star along the vertical cross-hair, as well as and his position and the precise bomb release point from the altitude of the star. *Sänger* said it would be unimportant whether or not the target area actually was visible to the pilot at the instant of bomb release.

A port side view of *Sänger's* personal *Dornier Do 17 Z-2* carrying one of his experimental ramjet engines. This was a tow test of a high-temperature ramjet engine of 1.6 feet [0.5 meters] in diameter and 18 feet [5.5 meters] in length. Gasoline combustion. Notice that, during operation, it is producing a steady, uninterrupted thrust...exactly what *Sänger* was seeking to perfect. During this airborne experiment its thrust coefficient was approximately 0.47 percent.

Another tow test, however, it appears that the thrust being produced by this particular *Sänger* experimental *Lorin* duct (ramjet) engine was less focused, therefore less satisfactory compared to the photo above.

A close-up starboard side view of the mouth of one of *Eugen Sänger's* experimental *Lorin* ducts (ramjet) mounted on a *Dornier Do 17 Z-2* as it sits on the tarmac.

Bomb Destructiveness: Point Release Bombing Verses Area Release Bombing

Sänger wrote that his rocket bomber could be used to drop bombs for an point impact or an area impact. The process of impact for a "point" bomb release and an "area" bomb release differ fundamentally by having very different velocities and angles of impact. The ballistic procedures used in "point" bombing are well-known and documented and the same as those used for ordinary bombing, or heavy mortar fire. For example, in a dive bomb attack by the rocket bomber with either a 30-ton, 5-ton, or 1-ton bomb, having a final diving speed of 500, 300, or 260 meters/second [1,640, 984, or 853 feet/second respectively] the penetrating power of the projectile through the earth's crust would be 100, 30, or 12 meters [328, 98, and 40 feet respectively]. For penetration power of a bomb through reinforced concrete...the values would be only 10 percent of the above numbers. *Eugen Sänger* thought his rocket bomber would be most effectively used for ultra-long range bombing attacks. The

***Sänger's* personal *Dornier Do 217 E-2* carrying an experimental 20,000 pound thrust *Sänger* ramjet engine as seen from its rear starboard side. This engine has a diameter of 4 1/2 feet [1.5 meters] and an overall length of 34 feet 7 inches [10.6 meters]. Notice that it has a dorsal fin, too, which is about 1/3 the length of the ramjet engine. *DFS*-Ainring Summer 1944.**

A poor quality photo of *Sänger's* experimental 20,000 pound thrust ramjet engine as seen from the *Dornier Do 217 E-2's* nose port side. *DFS*-Ainring Summer 1944.

A poor quality photo of *Sänger's* personal *Dornier Do 217 E-2* with the 20,000 pound thrust *Sänger* experiment ramjet as seen from its nose starboard side. *DFS*-Ainring Summer 1944.

type of bombing attack procedure he viewed appropriate for his rocket bomber would have been determined by the strategic importance of the target and its distance from the bomber's home airbase.

A Sänger Rocket-Powered Transport – the Beginning of the Concept

The idea for a rocket-powered vehicle first began in the early 1930s with the idea of developing an advanced commercial transport aircraft for carrying passengers. The shape of this rocket-powered transport airplane, *Sänger* wrote, would be similar to that of a bullet coming out of a rifle barrel:

"One will choose for the body the shape of a projectile, pointed in front and blunt at the rear to give room for the exhaust velocity. The profile of the wings has to be as thin as possible, with sharp leading edges. The wing span can then be kept low because of the

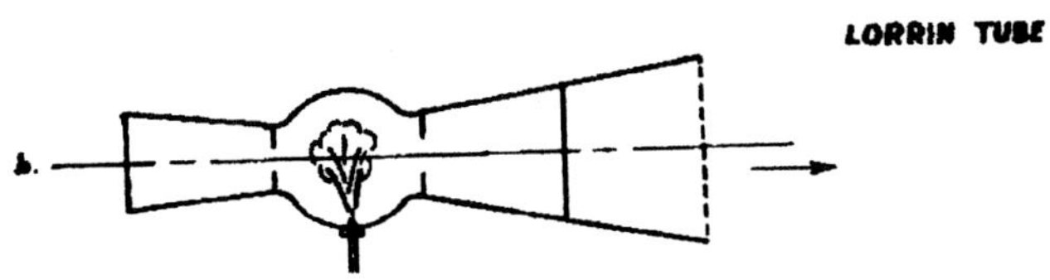

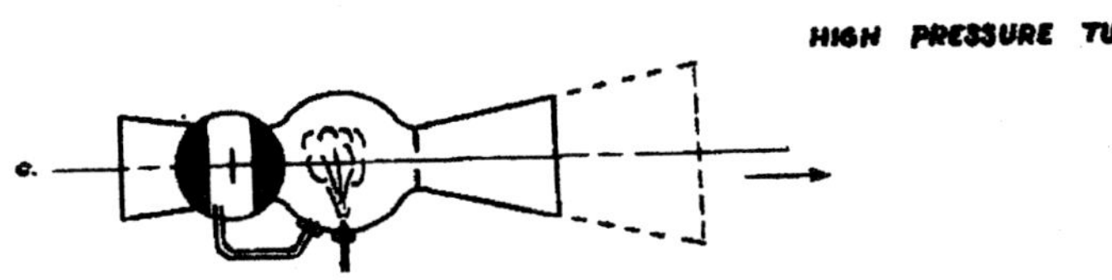

A pen and ink drawing featuring the internal details of two high-speed *Lorin* duct (ramjet) engines *Eugen Sänger* was seeking to perfect. While he was at *Professor Georgii's DFS*-Ainring research center, *Sänger* was informed that the ultimate purpose of his high speed *Lorin* duct research was to perfect an engine which could be used as an auxiliary engine on *Wernher von Bruan's* piloted *A-9* rocket missile with its *A-10* booster...*Wernher von Bruan's Amerika Bomber* project.

A poor quality photo of *Sänger's* personal *Dornier Do 217 E-2* in a high-speed dive with its *Sänger's* 20,000 pound thrust experimental ramjet in full working order at *DFS*-Ainring. Notice the length of the flame behind the *Dornier Do 217 E-2*.

Sänger's personal *Dornier Do 217 E-2* (*RE+CD*) seem during a flight test with an experimental high thrust *Lorin* duct (ramjet). Flights like the one shown here was conducted at the airfield at Hörsching between September 16th 1943 and August 30th 1944.

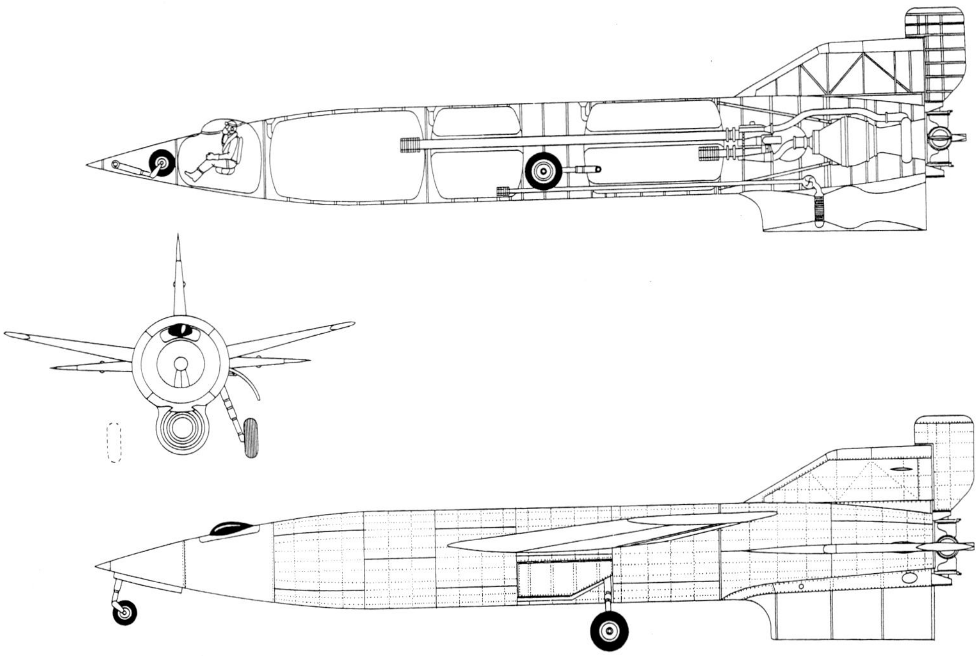

A pen and ink illustration of *Wernher von Braun's Eugen Sänger*-equipped *Lorin* duct auxiliary engine mounted dorsally beneath the vertical rudder. Courtesy: *Justin Miranda, The Reichdreams, V.T.O. Interceptor*, Dossier #9.

A *Wernher von Braun* piloted *A-9* rocket missile bomber's landing run is seen being slowed by its drag parachute. Digital image by *Josha Hildwine*.

A starboard side view of *Wernher von Bruan's* proposed piloted *A-9 Amerika Bomber.* In addition to its liquid rocket motor, this flying machine would have also featured an auxiliary *Lorin* duct (ramjet) engine from *Eugen Sänger.*

negligible resistance of the wing-edges." [A 90-foot wing span would probably be too broad, when considering this span is roughly equal to the fuselage length and would make *Sänger's* proposed vehicle quite unstable at the *Mach* levels anticipated at take-off and cruise.]

The anticipated forward velocity of the rocket-powered transport rocket-powered airplane, *Sänger* wrote, might be as fast as 29,000 kilometers/hour [18,021 miles/hour]:

"At first sight, their limit is given merely by the possible fuel load. It is this restricted fuel load which prevents these planes from increasing their flight velocity up to the orbital velocity of about 29,000 kilometers/hour. When the centrifugal force of the curved trajectory is equal to the weight of the plane; the wings then don't need to provide lift anymore, and the plane circles the Earth continuously, like a moon in a free orbit without needing any driving power."

A poor quality photograph of *Focke-Wulf Flugzeugbau's* ramjet specialist *Dipl.-Ing. Otto Ernst Pabst*, Bad Eilson, Germany. Summer 1943.

Three-view images of *Wernher von Braun's* proposed *Amerika Bomber* featuring an auxiliary dorsal mounted *Lorin* duct (ramjet) engine by *Eugen Sänger*. Digital image by Josh Hildwine.

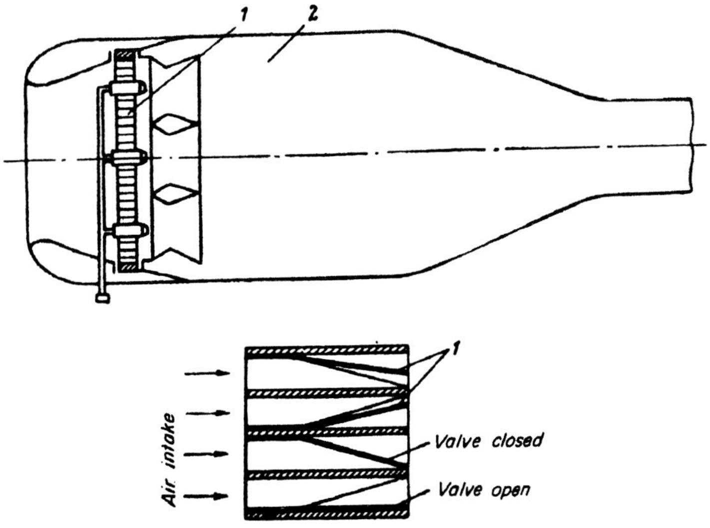

***Otto Pabst* had followed *Eugen Sänger's* ramjet engine experiments with great interest. He believed that they didn't need to have the long length *Sänger* preferred. *Pabst* began experimenting with much shorter length ramjet engines and made many combustion tests with his ramjet engines burning gasoline as fuel. Gasoline was vaporized and superheated outside the ramjet engine before being fed into the burners. *Pabst* accomplished this by building a heat exchanger into the downstream cross-section of his very small ramjet engines. For an aircraft installation requiring two ramjet engines each having 50 burners, it required only one small heat exchanger-ramjet with two burners.**

***Otto Pabst* was hoping to equip the three-bladed propeller powered *Focke-Wulf Fw Triebflügel* with a ramjet engine at each of the three propeller tips. The proposed propeller assembly had not yet been tested by the time of Germany's surrender. Seen are three *Focke-Wulf Triebflügels* as they might have appeared flying in formation. Digital image by *Mario Merino*.**

The *Otto Pabst* ramjet engine equipped proposed *Focke-Wulf Fw Triebflügel* vertical takeoff machine. *Pabst's* ramjet engines were shorter [500 mm] compared to their diameter [250 mm] than either of the *Walter HWK* or *Sänger* designs. *Walter's HWK* experimental ramjet had a length of 1.75 meters and a diameter of 250 mm. Scale model and photograph by *Günter Sengfelder*.

In the preface of his 1933, book *Raketenflugtechnik* or "The Technique of Rocket Flight," *Eugen Sänger* presented the details and flight performance of his rocket-powered transport airplane more extensively than he had done up to that time.

"In a more limited sense the project deals with that of a rocket flight in the upper layers of the stratosphere, with such velocity that the forces of inertia of the orbit contribute to the lifting effect. This kind of rocket flight is the following fundamental step in the phase of development from the troposphere flight established during the last thirty years. It is the preliminary stage of space flight, the most powerful technical problem of our time. This preliminary stage, and the development to the construction of an orbiting space station of the Earth, is the most noble aim of rocket flight, though its realization still lies in the future".

Sänger had considerable difficulty in getting his manuscript for his *Raketenflugtechnik* accepted by a publisher, yet today it is regarded as the first real textbook on rocket flight engineering. *Sänger's* book covered the groundwork on rocket combustion chamber testing. His interest in rockets grew as a natural consequence of his studies of aeronautics, where he quickly realized that high alti-

The *Focke-Wulf Fw Ta 283*. This proposed fighter by *Fw's* design genius *Hans Multhopp* would have had twin *Pabst* ramjet engines. Takeoff up to ramjet operating speed would have been provided by a *Walter HWK* bi-fuel liquid rocket engine installed within each tail-mounted *Pabst* ramjet engine. Scale model and photograph by *Günter Sengfelder*.

Another *Focke-Wulf* proposed day fighter design by *Hans Multhopp* and featuring two *Pabst* ramjet engines. Scale model and photograph by *Reinhard Roeser*.

tude flight resulted in an ever-increasing decline in the efficiencies of conventional propeller-driven aircraft. These inefficiencies would be overcome by the use of rocket propulsion, allowing flights to be made over vastly greater distances and at far higher altitudes when compared to conventional aircraft using internal combustion engines. *Sänger* began a detailed theoretical examination of the basis of rocket propulsion, ranging from the use of combustion chamber materials such as aluminum, beryllium, and lithium. He investigated various propellent agents and sources of oxygen, including nitric acid, hydrogen peroxide, and liquid oxygen.

Between 1930 and 1933, courtesy of the Vienna, Austria *Technische Hochscule* or "Technical High School," *Sänger* successfully conducted numerous experiments with rocket combustion chambers in two small wooden sheds on the school grounds. It was a modest installation consisting of a small workshop in one shed containing tools, equipment, and a pump room lined with pressure

An original *Sänger-Bredt* design for a fighter powered by a single 60,000 horsepower ramjet engine. This *Sänger-Bredt* design was turned over to *Skoda-Kauba*-Cakowitz (near Prague) for development, and it became known as the *Skoda-Kauba Sk P.14-01*. It is shown here from its port side nose. Digital image by *Josha Hildwine*.

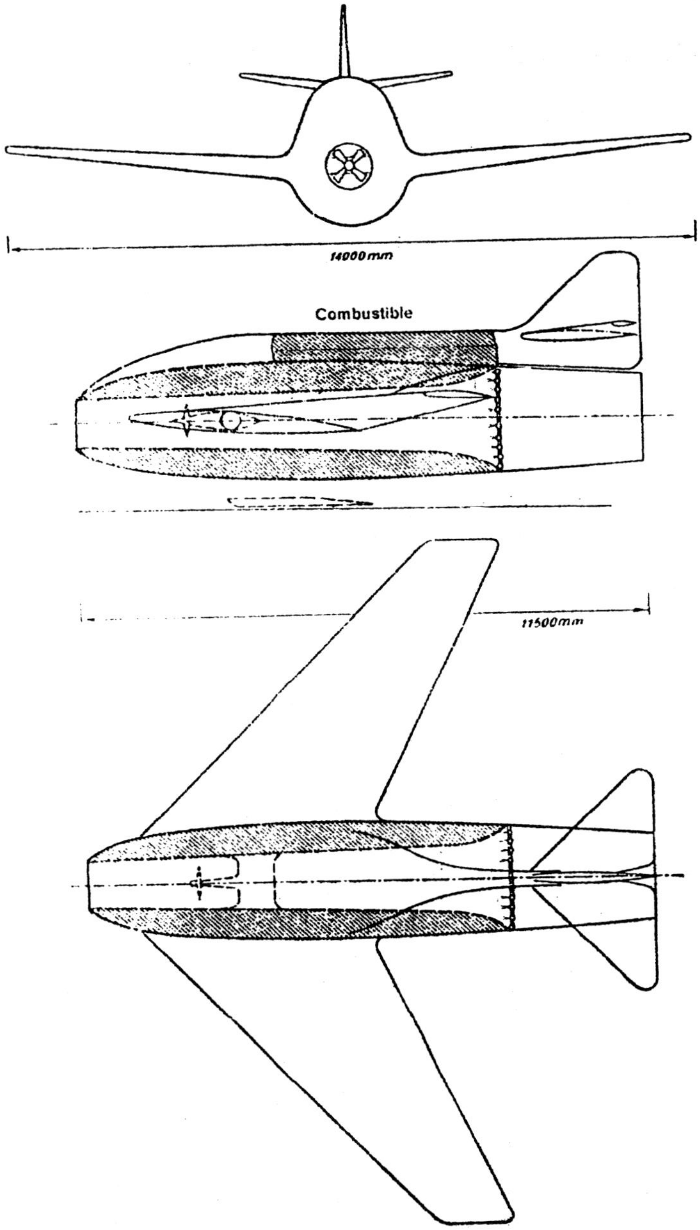

A pen and ink three-view illustration from *Skoda-Kauba* featuring their proposed *Sänger-Bredt* powered ramjet-rocket engine combination fighter, the *Skoda-Kauba Sk P.14-01*:

- Role - **fighter;**
- Engine - **1x60,000 horsepower *Sänger-Bredt* ramjet;**
- Takeoff assistance - **not available;**
- Fuel capacity - **212 gallons;**
- Wing span - **23 feet [7.00 meters];**
- Length - **32.3 feet [9.85 meters];**
- Speed, sea level - **621 mph;**
- Speed, cruising - **398 mph at 42,650 feet;**
- Ceiling - **60,700 feet;**
- Armament - **1x*MK 103* cannon in the nose;**
- Crew - **one, piloting the aircraft from a prone position;**

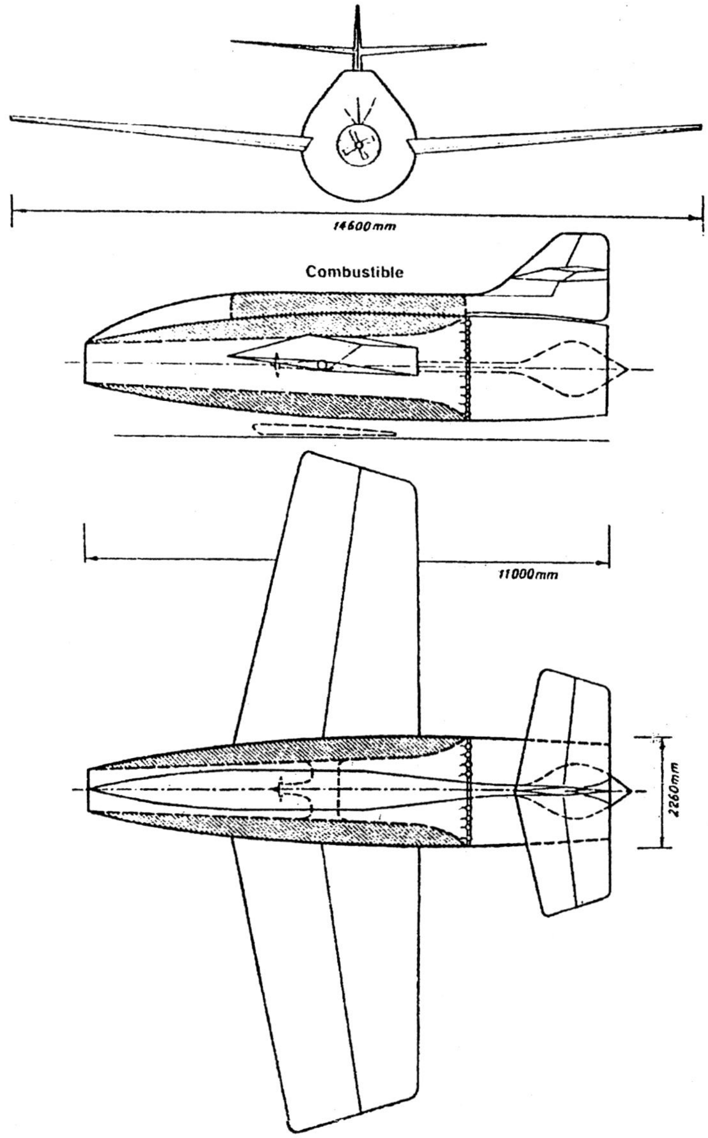

A pen and ink 3-view drawing of the *Sänger-Bredt* ramjet-rocket combination powered *Skoda-Kauba Sk P.14-02*. This proposed ramjet-engined fighter was similar to the *Sk P.14-01* but with only minor modifications, especially the cockpit canopy.

gauge dials. The second shed, open on one end, held a metal framework test stand for mounting various types of combustion chambers being tested. Each test was controlled from the workshop shed — the so-called "pump room" — where *Sänger* could monitor pressure gauges while he also observed the actual test firing of the combustion chamber in the adjoining shed through a vision slit in the wooden wall. Most of these rocket combustion chambers were quite small— small enough to be held in the palm of one's hand. However, each one possessed all the characteristic features of much larger conceptual designs — injection nozzles, a shaped combustion chamber, and a regenerative cooling subsystem. With virtually no outside financial or technical help, *Eugen Sänger* methodically pursued experiments on constant-pressure combustion until he had

A *Sänger-Bredt* ramjet-rocket combination powered *Skoda-Kauba Sk P.14-02* is shown shortly after lift-off from its tricycle dolly. Digital image by *Josha Hildwine*.

enough experience and data to compile a comprehensive study of practical rocket propulsion technique. Every experiment was meticulously noted in a 300-page daily logbook. Many of these entries were then included in his subsequent book manuscript. In December, 1933, *Sänger* submitted his work to the Austrian Federal Defense Ministry seeking funding for further experiments dealing with rocket-driven aircraft and military artillery projectiles. On February 3rd 1934, *Sänger* received this reply:

"In reply to your letter of December 26th 1933, you are informed that the Federal Defense Ministry is unable to give further consideration to your rocket project since its basic principle (the use of liquid hydrocarbons and liquid oxygen) would be incapable of practical realization, due to the excessive danger of explosion inseparable from such propellents."

A group of three *Sänger-Bredt* ram-jet rocket combination powered *Skoda-Kauba Sk P.14-02* fighters engage a formation of *Boeing B-29's*. We see one of the *B-29s* moments after it was hit. One of the *Sk P.14-02s* has been hit, too, trailing smoke and flames. Digital image by *Josha Hildwine*.

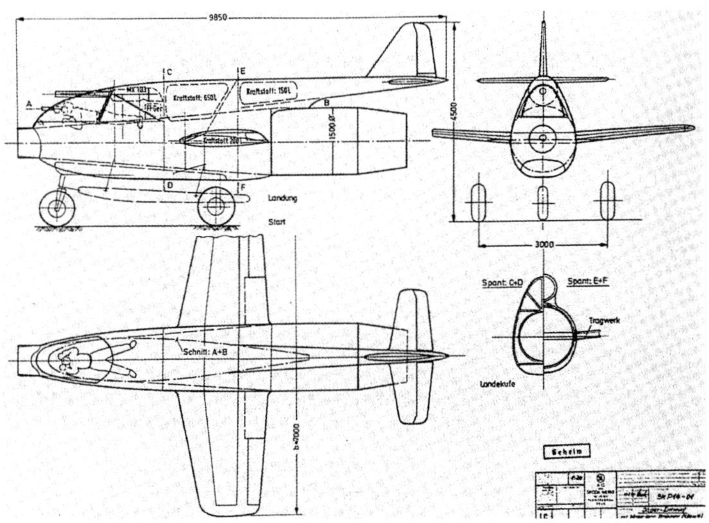

***Skoda-Kauba* proposed several other day fighters, each propelled by a ramjet-rocket combination *Sänger- Bredt* engine. One of their proposed day fighters was known as the *Skoda-Kauba Sk P.14-04* shown in this 3- view drawing.**

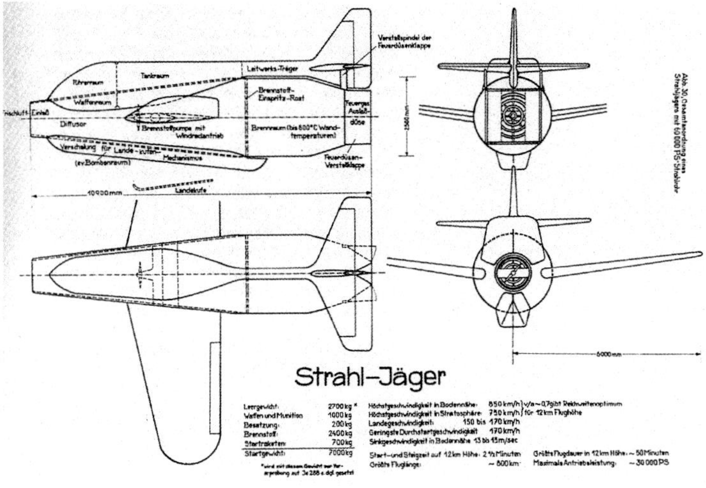

A pen and ink 3-view drawing of a proposed *Skoda-Kauba* miniature day fighter to be powered by a *Sänger-Bredt* 60,000 (*PS*) horsepower ramjet-rocket composite engine. It wasn't until November 30th 1944, that the *RLM* gave its permission to construct a prototype ramjet-rocket composite engine.

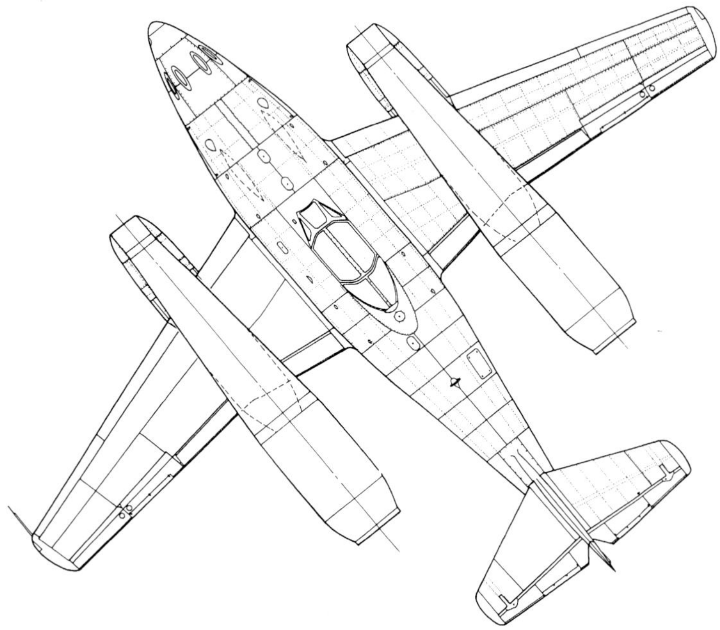

A pen and ink overhead plan view of the proposed *Eugen Sänger* ramjet engine equipped *Messerschmitt Me 262A* twin turbojet powered fighter.

A *Messerschmitt Me 262A* shown being flight-testing with two *Eugen Sänger* ramjet engines in addition to its two *Junkers Jumo 004s* turbojet units. Digital image by *Mario Merino*.

The *Eugen Sänger* ramjet engine on a *Messerschmitt Me 262A* as seen from its rear port side. Their size dwarfs the *Junkers Jumo 004Bs*. Digital image by *Mario Merino*.

The *Eugen Sänger* ramjet engine-equipped *Messerschmitt Me 262A* as seen from its starboard side. Digital image by *Mario Merino*.

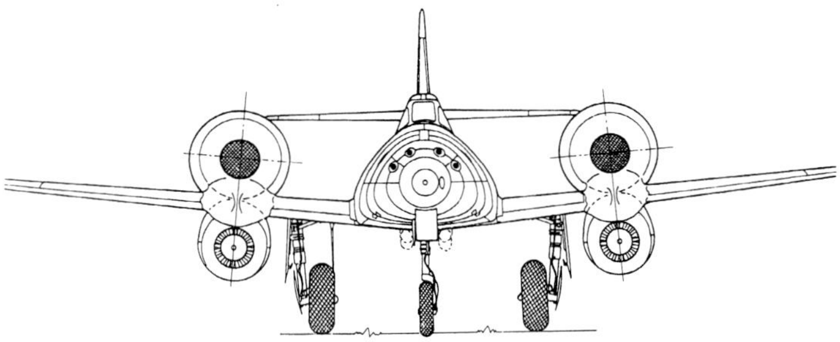

A pen and ink drawing of how the proposed *Messerschmitt Me 262 Eugen Sänger* ramjet engine equipped fighter would appear as seen head on.

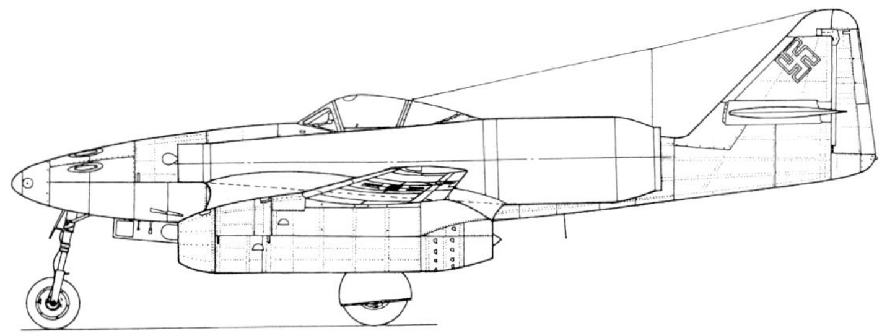

A pen and ink port side drawing of the proposed *Eugen Sänger* ramjet powered *Messerschmitt Me 262* as seen from its starboard side.

As for his book manuscript, by the Summer of 1932, *Sänger* had received 11 rejection letters from various publishers. Finally, a small publisher in Berlin, who had previously published *Hermann Oberth*, *Walter Hohmann*, and *Max Valier's* books, agreed to print *Sänger's* rocket manuscript, provided that he contribute 2,000 *Reichmarks* toward its printing costs. In addition, *Sänger* had to agree to purchase the first 50 copies printed. It was not until 1936, when he was hired by *DVL* as the technical director of their fledgling rocket research program, that *Sänger* was able to pay off his debts associated with the printing of *Raketenflugtechnik*.

About a year after his book was published – in the Summer of 1933 – *Sänger* outlined a concept for a development and testing program for his proposed rocket-powered transport airplane in the journal *Flug*. Reading his carefully kept log-book, one sees advanced and innovative technical concepts still in use today, such as:

- The combustion chamber may be cooled by the use of the fuel or the oxidizing agent (regenerative cooling);
- High-pressure oil cooling of the chamber is possible;
- The transmission of heat is independent of the degree of mechanical efficiency;
- Highly efficient utilization of propellants is obtained from exhaust jets with a wide-angle aperture;
- Light metals can actually be used as a fuel;

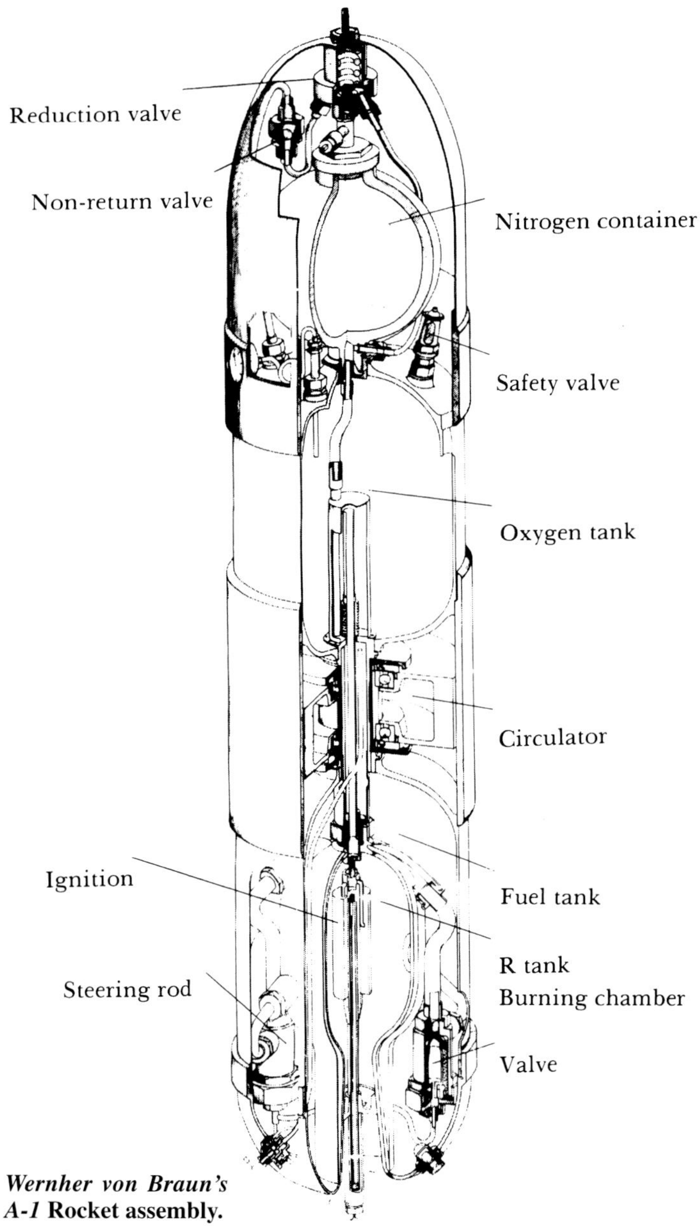

***Wernher von Braun's A-1* Rocket assembly.**

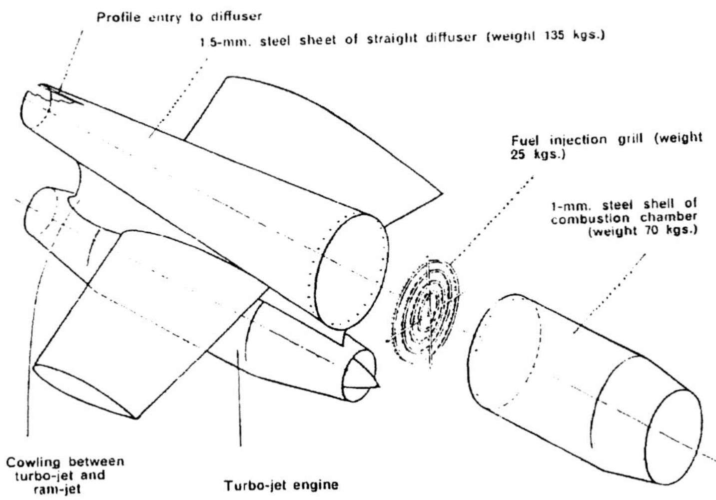

The main components of *Eugen Sänger's* experimental *Lorin* ramjet engine fueled by heating oil as installed on a *Messerschmitt Me 262A*.

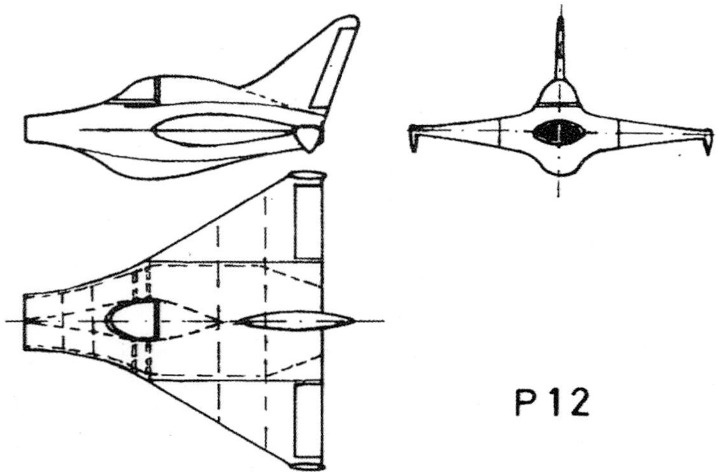

The *Alexander Lippisch* ramjet engine-powered single seat fighter ramjet design known as the *Lippisch Li P. 12. Johannes Winkler* of *Luftfahrt Forschungs Anstalt (LFA)*. This and other *Lippisch* ramjet designs were to use solid carbon for fuel instead of liquid carbons. *Winkler* was considered the leading researcher in solid carbon-powered ramjet engines. It is not entirely clear, however, it appears that *Eugen Sänger* was not involved in solid carbon fuel for ramjet engines.

- When constant ignition is maintained by minute particles of liquid oxygen and vaporizing oil, combustion will occur in precisely the same way as when gaseous oxygen is used;
- A mixture of liquid oxygen and oil droplets will burn evenly and rapidly without exploding (regardless of the what the scientists at the Austrian Ministry of Defense believed).

With a growing list of successful experiments, *Sänger's* long range plans were to increase thrust to 60 pounds from the 20 pounds he was obtaining. With 60 pounds of thrust, exhaust velocities would reach 10,000 feet/second. Next he would fabricate a combustion chamber producing 100 pounds of thrust, with an ultimate goal of developing a combustion chamber capable of 1-ton (2,000 pounds) of thrust which could be tested in an aircraft. *Sänger's* dream was the construction of a high-altitude aircraft capable of reaching the troposphere and achieving a level flight speed of 600 miles/hour. He believed that fighters and bombers could be constructed with his rocket motors allowing them a range of 3,000 miles. Finally, he wanted to construct a flying machine which could circle the globe and then go on to build a space station for earth orbit, which could be used as a 'factory' for construction of vehicles for interplanetary flight.

Sänger conducted over 200 practical tests with his rocket combustion chambers. One experiment produced an exhaust velocity of 8,300 feet per second and combustion gas temperatures reaching 3,000 degrees Centigrade. *Sänger's* combustion chamber experiment results with his test *Model 14* produced an exhaust velocity of 9,750 feet/second, a pressure of 60 atmospheres, and a thrust of 65 pounds. But *Sänger* was told by the *Technische Hochscule's* direc-

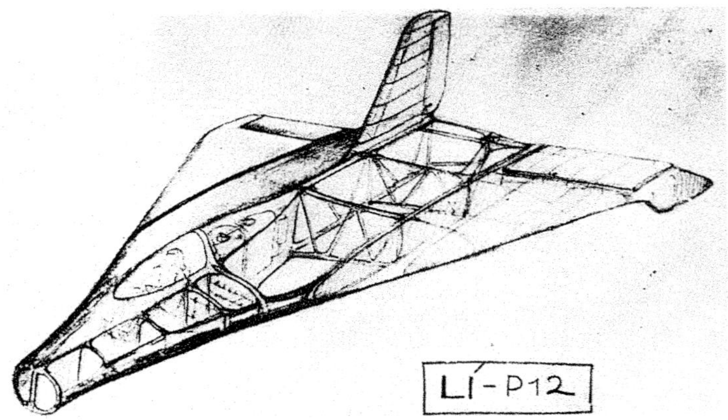

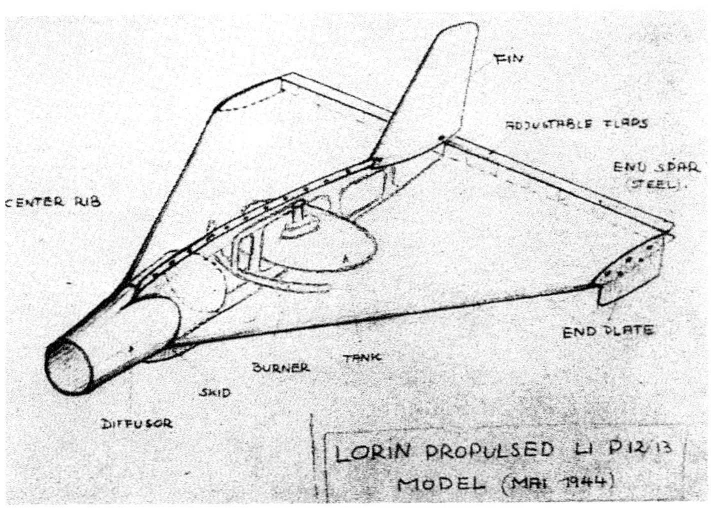

A pen and ink see through illustration of *Alexander Lippisch's* solid carbon fueled ramjet powered *Lippisch Li P.12* from March 1944. The lower illustration shows the shape of *Johannes Winkler's* circular container or basket for holding its solid carbon fuel.

***Alexander Lippisch* seen in one of his rare happy moods. Left to right: *Hans Jacobs*, *Lippisch*, and *Wilfried Teichmann*. *Lippisch* favored *Johannes Winkler's (LFA)* solid fuel ramjet engines over those of *Eugen Sänger's (DFS)* liquid carbon fuels.**

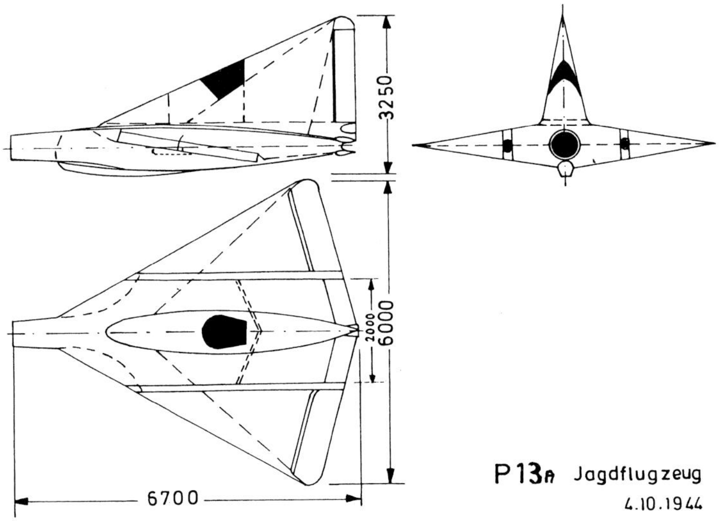

***Johannes Winkler's (LFA)* solid carbon ramjet engines varied extensively of those being researched by *Eugen Sänger* at *DFS*. *Sänger's* ramjets were powered by liquid carbons and their burners were housed in a long tube. *Winkler's* ramjet was more rectangular, in many cases round. Shown is a pen and ink drawing of the proposed *Alexander Lippisch Li P.13a* solid fuel ramjet fighter prototype from October 4th 1944.**

tor to bring an end to his rocket combustion chamber testing because neighbors were complaining of noise.

In a special edition of *Flug* printed a year later in the Summer 1934, *Sänger* presented a theoretical aircraft's flight performance and flight trajectories based on a fuselage with circular cone-shaped nose, flat and extremely thin wings, a mean lift-drag ratio of 05.0, and an effective exhaust thrust velocity of 3,700 kilometers/hour [2,300 miles/hour]. He wrote that for a mass ratio of *G/Go* of 0.16 (the ratio of empty weight to gross take-off weight), for example, the velocity at burnout would be about *Mach 13* or about 8,500 miles/hour, a steady flight velocity of *Mach 3.5* at an altitude of 40 to 60 kilometers [196,900 feet], and a flight range of 5,000 kilometers [3,100 miles]. *Sänger* wrote:

"Accordingly, the most important task of the (airplane) constructor has to be the achievement of an adequate range by severe decreases of empty weight, by the most sophisticated aerodynamic shapes, and by the highest possible exhaust velocity of the engine. But even with the not unrealistic assumptions so far used, a rocket plane with a non-stop range of 4,000 to 5,000 kilometers can be confidently expected."

Following *Sänger's* extensive presentations in journal articles and newspaper essays during 1933, not much time was devoted to his civilian orbital rocket-powered transport airplane. Available hours were limited to testing small prototype liquid rocket propulsion engines of very low thrust which *Sänger* himself designed and built. He wrote that before even constructing or testing the rocket plane's airframe, its engine's thrust would have to be carefully calculated. So why even consider how the airframe would look?

Engine experiments were carried out while *Sänger* was still a scientific assistant at the *Technische Hochscule* in Vienna and these

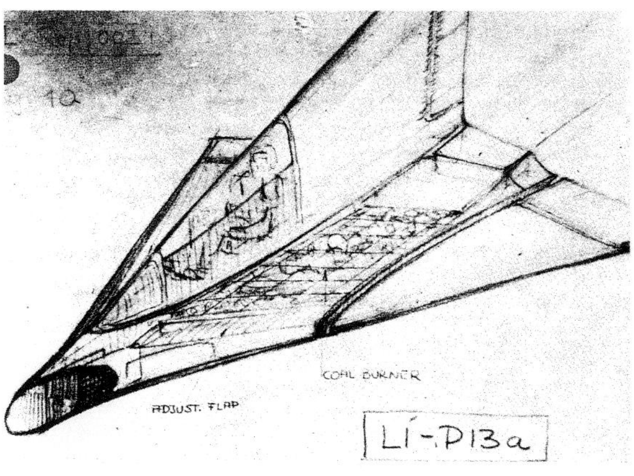

The proposed solid carbon fueled ramjet engine fighter design of *Alexander Lippisch* and known as the *Lippisch Li P.13a* and patterned after his *DM-1* glider. The machine would require an auxiliary power unit for take off as the *Johannes Winkler* ramjet engine required a forward speed of at least 149 miles/hour [240 kilometers/hour] before it would become self sufficient.

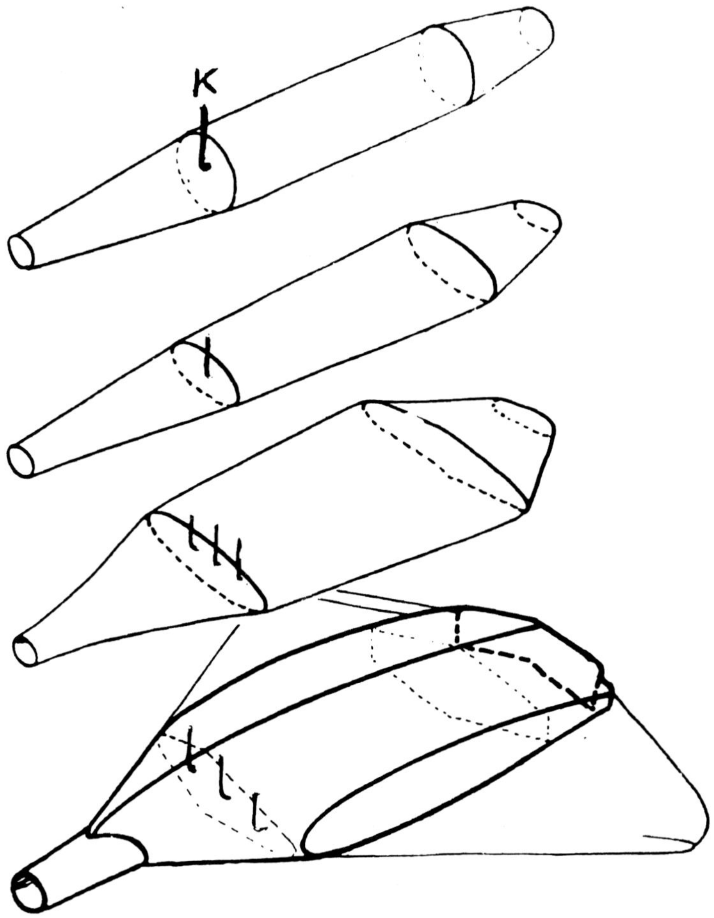

A pen and ink illustration showing the evolution of the ramjet engine. At the top is a typical *Eugen Sänger* liquid fueled ramjet. "*K*" is the fuel nozzle. *Johannes Winkler* carried the ramjet design further from its stove pipe style favored by *Eugen Sänger* into a more rectangular shape and adding up to three fuel nozzles.

tests continued up to the end of 1934. They were performed at the *Institute für Baustoffdkunde* (commonly known as the "*Bauhof*") or Institute for research into building materials. This testing was done exclusively on liquid propulsion combustion systems utilizing regenerative cooling. It was the first practical application of this principle anywhere and *Sänger*, with these prototype model engines, realized thrust velocities up to 3,000 meters/second [9,852 feet/second] during the testing.

By 1935, *Sänger* still had not found work in rocket research. Instead, he was supporting himself by working as a temporary engineer for a Viennese construction company, while he actively sought employment at rocket development facilities all over the world. It was a difficult time for him. Then, on February 1st 1936, *Sänger,* found employment with a rocket research institute. It was with the *Hermann Göring Institute,* the research arm of the German *Luftwaffe*. The parent of this Institute was the *Deutsch Versuchsanstalt für Luftfahrt*-Berlin (*DVL*), or German Research Establishment for Aviation. His contract must have been very pleasing to him because *DVL* wanted to establish a secret rocket research

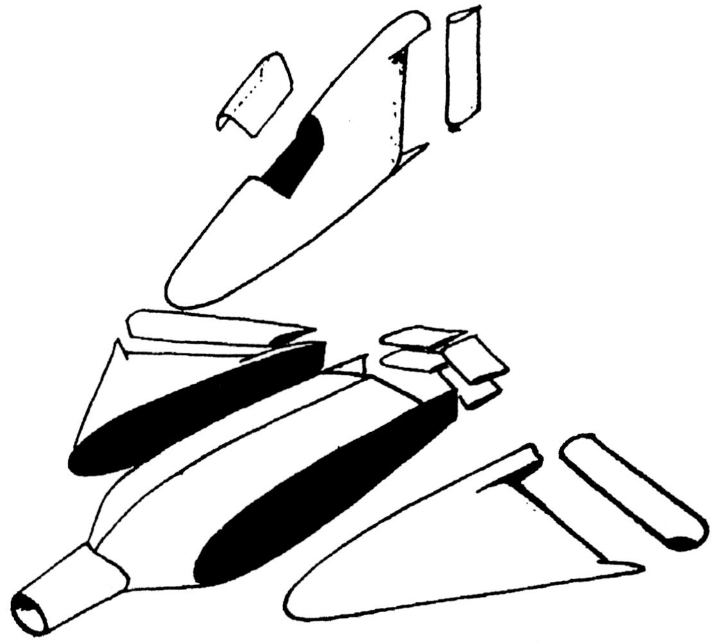

A pen and ink illustration of the several component parts to the *Alexander Lippisch* designed single seat *Lippisch Li P. 13a* to be powered by a solid carbon fueled ramjet engine by *Johannes Winkler*. *Eugen Sänger* was not happy at all to be in a race with one of the early pioneers of rocket science...*Johannes Winkler* doing research on ramjet engines. He only wanted to be involved in research of his 100-ton thrust bi-fuel liquid rocket engine for his orbital bomber.

institute which would be separate from the German Army's research facilities at Peenemünde.

In reality, *DVL's* hiring of *Eugen Sänger,* Ph.D., who was at this time a young self-taught propulsive rocket researcher with several technical publications to his credit, came largely as a result of the German Army High Command's enthusiastic response from *Adolf Hitler* to their proposal for establishment of a rocket test cen-

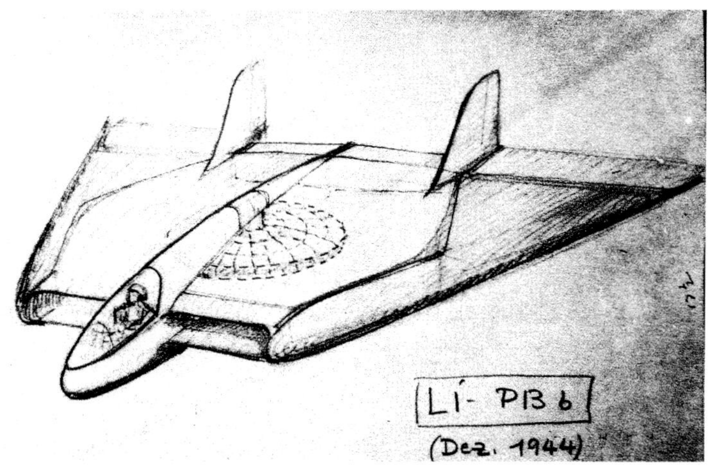

***Alexander Lippisch's* proposed solid coal dust ramjet enginend-powered *Li P. 13b* of December 1944. *Dr.-Ing. Johannes Winkler, LFA*-Braunschweig (Brunswick), was actively researching solid carbons such as lignite coal as fuel for ramjet engines. *Eugen Sänger* and *Johannes Winkler* of *LEA* were not friendly competitors.**

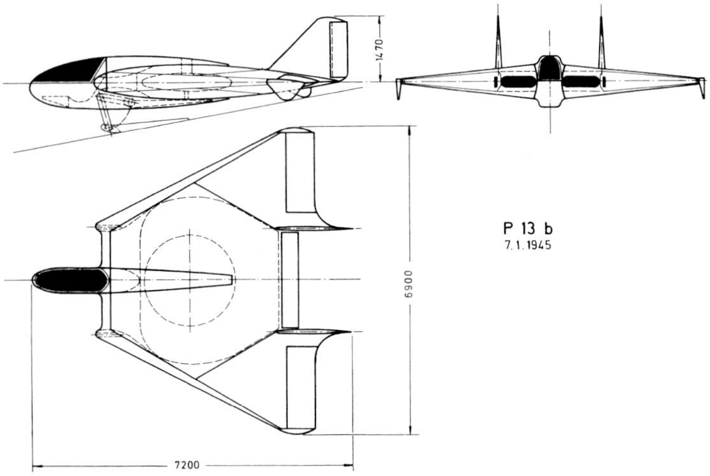

A pen and ink 3-view illustration of *Alexander Lippisch's Li P.13b. Johannes Winkler's* ramjet engine research favored the use of a round basket containing solid carbons. Ignition would be imitated by a propane gas burner. The ramjet's thrust would exit an opening which occupied the center section's trailing edge between the two vertical rudders.

Former *Eugen Sänger* research assistant at *DVL* and friend *Helmuth von Zborowski's BMW 003R* rocket composite turbine with its bi-fuel liquid *BMW 718* rocket engine atop the standard *BMW 003* turbine. This power unit was suggested for a variety of *Luftwaffe* aircraft projects including the *Arado Ar 234C*...for extra thrust during take-off or rapid climbing during night-fighting action.

ter at Peenemünde Experimental Rocket Center. In 1935, a detailed memoir on the development of long-range, remotely controlled rocket bombs was submitted to *Hilter* by the Army Staff, under the authority of *General Dr. Karl Emil Becker*. In this report, the expectation was expressed that this new class of weapons could be developed rather quickly, provided that sufficient support and a large, special development establishment were made available under orders of the "*Heereswaffenamt*", or Army Armament Office. *Hilter* was delighted at the prospects offered, and the *Heereswaffenamt* obtained tens of millions of *Reichmarks* — with a promise of half that amount again in 1938. The lavishly-built rocket research and testing center at Peenemünde was the result of *Hilter's* direct intervention in spending decisions. Hearing of the Peenemünde Program, *Luftwaffe* boss *Hermann Göring* resolved not to be outdone by the *Wehrmacht* (the *German Armed Forces,* who were not friendly at all to *Göring* and his *Luftwaffe*) and urged his *Luftwaffe* planners to set up their own rocket development and research institute.

Sänger was asked by the *DVL* to design his own research facilities (institute) around a program involving liquid propellants and their testing in propulsion engines. The construction of facilities started in February, 1937, near the small villages of Trauen/Fassberg on the Lüneburg heath, which was accessible at that time only by horse and wagon. Just before the outbreak of *World War II*, the laboratory was ready to begin research. Officially, this facility came to be known as *DVL's Luftfahrtforschungsanstalt-Hermann Göring* (*DVL/LFA*) or "The *Herman Göring* Research Institute for

Count von Helmuth Zborowski.

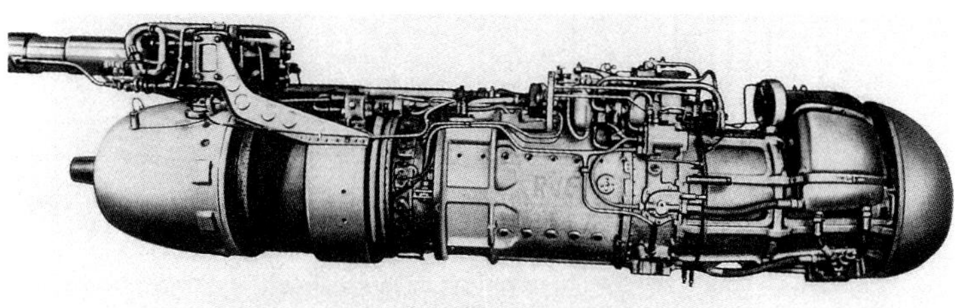

A starbroad side view of *Helmuth Zborowski's BMW 003R* combination bi-fuel rocket engine and turbojet engine.

The port side tail of the *Messerschmitt Me 262 V170074* shown being static tested with its *Zborowski*- designed *BMW 003R*. Notice the ferocity and length of the flame emitting from *Zborowski's BMW 718* bi-fuel liquid rocket engine as it covers the hangar concrete floor. The flame and noise had to be scarey for everyone involved.

A close-up view of the *BMW 718* by *Helmuth von Zborowski* installed on the *Messerschmitt Me 262 V170074* with its blow-torch like white hot flame. The port side of the *V170074* is shown.

Starboard side view of an experimental *Zborowski* bi-liquid rocket engine on its test stand.

Forward starboard side view of an experimental *Zborowski* bi-liquid rocket engine on its test stand.

Rear view of the experimental *Zborowski* bi-liquid rocket engine after it blew up during ignition.

Rear view of an experimental *Zborowski* bi-liquid rocket engine ready for its initial test firing.

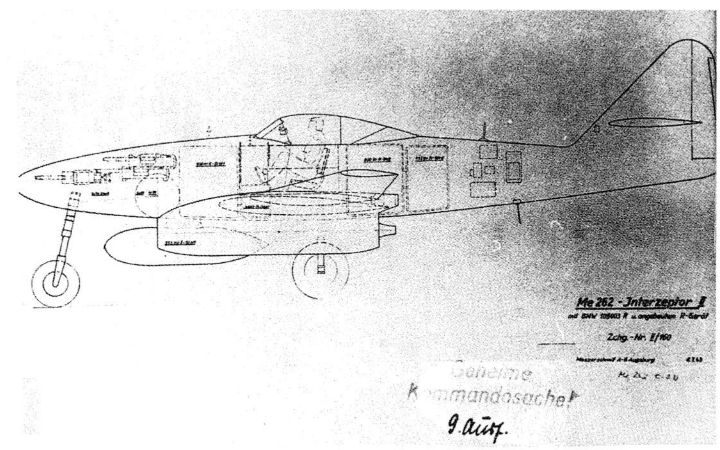

A pen and ink illustration by *Messerschmitt AG* of the *Zborowski*-designed *BMW 003R* as installed in a *Messerschmitt Me 262* for testing. Viewed from its port side.

Aviation". Money subsequently invested into rocket research by the *RLM* seemed to be unlimited; it's estimated that the *Luftwaffe* may have poured as much as 8 million *Reichsmarks* into their "Aircraft Examination Point" facility. For example, the Trauen/Fassberg facility included the largest liquid-oxygen (lox) container ever built, capable of holding 110,000 pounds of the liquified gas. *Sänger's* program provided for ten years' fundamental research on liquid-fuel rocket engines, after which engines should become available for long-range aircraft and – in particular – for long-range bombers. *DVL's* aim was to design and fabricate a rocket engine producing 100-tons (200,000 pounds) of thrust for an extended time period. Rocket motor mounting stands capable of testing these ultra high thrusts were built in anticipation of the day when a rocket engine of this magnitude would be test-fired. It is interesting to note here that, while the Army only considered development of ballistic missile type projectiles such as the *V2* rocket missile of *Wernher von Braun*, *Sänger* gave preference to a more scientifically-oriented general approach to rocket-powered winged aircraft and missiles. He was planning to conduct research leading eventually to supersonic flight, aimed at speeds between *Mach 3* (1,863 miles/hour) and *Mach 30* (18,630 miles/hour) at extremely high altitudes. *Sänger* was saying that at a height of 80 miles altitude, a rocket-powered flying machine would lose about 6 miles of altitude per revolution about the Earth as a consequence of air drag. But if the vehicle could circle the Earth at a height of 112 miles, this altitude loss per revolution would be only 3 feet because of the near total absence of friction-inducing atmospheric drag at that level. *Sänger* believed that air resistance and resulting skin heating for aircraft in the range of altitudes between 80 and 112 miles altitude would reduce forward velocities and effective range considerably.

The *Zborowski*-powered *BMW 003R Messerschmitt Me 262 V074* as seen from its starboard side. Scale model and photograph by *Jamie Davies*.

Looking back to this time, it appears that *Sänger's* long term rocket research program was more impressive than the short-term, over-optimistic policy of the Army researchers at the Peenemünde Experimental Rocket Center, which had been based on *Hermann Oberth's* concepts for space flight.

Sänger's ideas for an orbital transport and orbital bomber also appear to have been influenced by *Oberth*, who was a *Romanian* grade-school teacher. In 1927, *Oberth* wrote that a rocket-propelled airplane with an empty weight of 8,850 pounds would require about 53,000 pounds of fuel to fly a distance of 1,050 miles. *Oberth* concluded that flight trajectories at extreme altitudes, followed by glides at lower altitudes, would provide a more economical means of covering extensive distances. This was the idea *Sänger* advocated in the early 1930s for his orbital civilian transport aircraft, and then in 1942 for his proposed orbital winged rocket bomber. Only in 1944, however, was a condensed version of *Sänger's* report prepared and a few copies circulated.

LFA Institute Hermann Göring-Trauen/Fassberg

Eugen Sänger and his two research assistants, *Dr.-Physics Irene Bredt* and *Dipl.-Ing. Count Helmuth von Zborowbski,* worked at *LFA* between August, 1937, and August, 1942. The main program was research and development work on *Sänger's* 100-ton prototype liquid propellant rocket engine design. *Sänger* said that his rocket research work was delayed several years due to the constructing of the Trauen Aeronautical Testing Station, however, by 1939, his testing again resumed. The construction of the experimental installations was under the direction of *Sänger's* research assistant *Helmuth von Zborowski*. The construction of the components was directed by *H. Ziebland* and *K. Hedfeld* directed the experimental work. *Irene Bredt* was twenty-five years old when she completed her doctoral degree in physics at the University of Bonn in 1936. Her advisors were *Professors Konen* and *Fuchthauer*. Her doctoral thesis was on "X-ray Spectra of Rare Earths." Following her graduation she received four offers of employment. Stimulated by the secret (rocket) military work at the small village of Trauen/Fassberg, she selected this offer over the others and was hired to be *Dr.-Ing. Eugen Sänger's* assistant. *Eugen Sänger* was a dreamer, shy, charming, six years older than *Irene Bredt,* and his ultimate goal was to create a manned rocket-powered vessel which could reach space. Although it may have been *Opel*, it is not known who in Germany first coined the word "*rakete*", from the Italian word "*roccketta*", meaning the mediaeval gunpowder propelled fire arrows. In English, it became "rocket." In 1897, German novelist *Kurd Lasswitz* published a novel titled "*Auf zwei Planeten.*" *Lasswitz* wrote that his space ship was stopped, started, and steered by "*repulsor*" engines which created a regulated explosion. The term "*repulsor*" later became favored among the experimenters of the German Rocket Society as a general term meaning rocket engine.

The Soviets Search for *Sänger* to Give Them a New Cold War Advantage

When the Soviet spy (and the USSR's chief Air Force rocket scientist) *Leutnant-Colonel Grigori A. Tokaev* defected to the West in 1947, it was finally revealed that the USSR already had full knowledge of the contents of the secret *Sänger* report. *Tokaev* later wrote about the details in his book *"Stalin Means War."* It appears that on April 14th 1945, there was a high-level meeting in the Kremlin in Moscow to discuss the technical feasibility of continuing development of the *Sänger* orbital bomber project in the Soviet Union.

***Professor Helmuth Walter*, founder of the *Walter HWK 509* bi-fuel liquid rocket firm located at Kiel.**

***Dr.-Med. Hubert Strughold* on his 89th birthday in 1986. He is considered the father of Space Flight Medicine.**

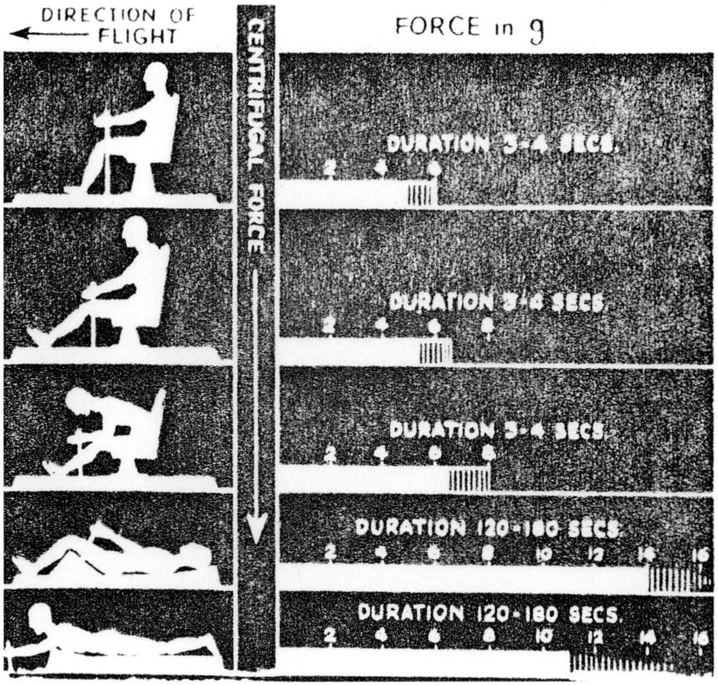

A poor quality pen and ink illustration featuring typical pilot seating positions and the amount of gravity (g's) a pilot siting can stand based on *Dr.-Med. Hubert Strughold's* research for the *Luftwaffe*. *Eugen Sänger* was planning to place his rocket bomber pilot in a position for takeoff somewhere between #3 and #4. Position #3 would allow most pilots to comfortably withstand up to 6+ g's. Position #4, with the pilot fully reclined, allowed most pilots to comfortably withstand to 11g's.

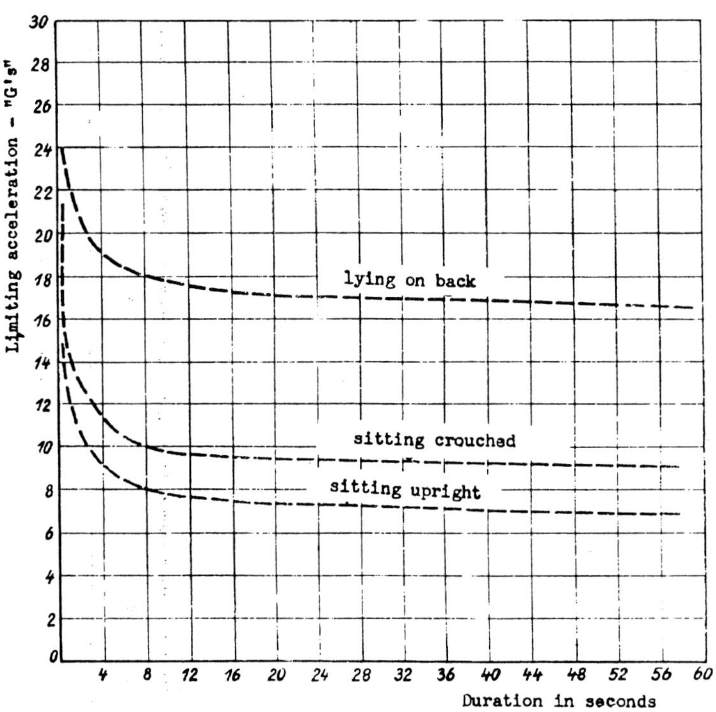

Tokaev was present at that meeting and was questioned extensively, because his immediate superiors were obviously very interested in the *Sänger* orbital bomber idea. *Tokaev* later wrote, after the end of *WWII*, that his opinion at the time was that *"Sänger* was a gifted scientist, but rather academic and lacking in practical experience."

"Only three copies of the *Sänger Project* were found among the piles of documents in: Berlin, Halle, and Peenemünde. The documentation consisted of a compilation of physical-mathematical, chemical, and gas-dynamics theoretical data for a super long-range, supersonic, stratospheric rocket-propelled bomber, capable of non-stop flight around the world. The first reaction of some Soviet scientists was to scoff at the idea, but after delving more deeply into the subject, they came to the conclusion that *"Sänger's"* bomber was feasible in principle. I, personally, came also to this opinion." [Tokaev]

Tokaev recalled cautioning his colleagues that *"Sänger's* physics and mathematics should be carefully checked." The very next day, *Tokaev* was called in to the presence of *Josef Stalin,* whose mind appeared to be already made up regarding support for the Russian counterpart of *Sänger's* orbital bomber project. The result of that meeting between *Stalin* and *Tokaev* was the formation of a Special Commission #2 for investigation of the *Sänger* project and production of a technical feasibility report by August, 1947. Several German rocket scientists who had been taken into the USSR post-war gave their technical opinions on *Sänger's* orbital bomber to *Tokaev*:

- They doubted *Sänger's* calculations of a mass ratio of 0.1 (the mass ratio of a rocket vehicle is its empty weight divided by the weight at launch). However, *Sänger's* theories were proven correct 20 years later when America's *Saturn 5* moon rocket had about the same mass ratio as he had predicted;

- The "skip" flight plan of the orbital bomber proposed by *Sänger* would produce severe loads on the aircraft's wing root structure; *Sänger* did not say how these loads were to be accommodated;

Left: A pen and ink drawing by *Eugen Sänger* featuring the limits of accelerations withstood by trained pilots in terms of "g" as a function of time of duration. This graph was the result of the most recent studies at that time (1944) on the highest accelerations that were believed to be bearable by a pilot in different body positions. It appeared that in the siting position the limit was determined by disturbance of the circulatory system especially costed by loss of blood from the brain or heart due to differences in hydrostatic level inside the circulatory system. This danger decreased in the lying position and the pilot's limit in the lying position seemed to be set by their difficulties in breathing as a result of the greatly increased weight of their thorax. *Sänger* estimated that the accelerations of his rocket bomber during takeoff would be about 10g's, well within the limit tolerated by a pilot in the lying position.

Left to right: *Professor Walter Georgii* and several members of his brood have gathered around a table for supper. The people present are individuals working on the *DFS 228* and *DFS 346.* The author is not aware of photographs showing *Walter Georgii* and *Eugen Sänger* together.

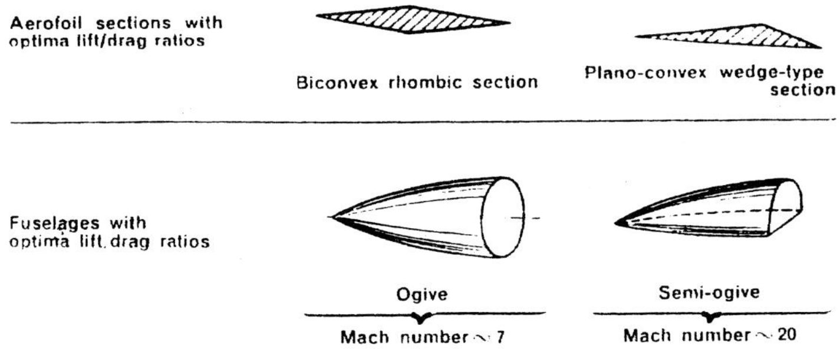

This graph presents the favorable flying-speed ranges of biconvex rhombic section and plano-convex flying wedge-type section bodies and fuselage noses with optima lift/drag ratios. *Eugen Sänger* discovered that at high *Mach* numbers, the pointed or cutting forward extremity of flying bodies must feature a form that the pressure side of the wing (which furnishes lift) is not curved in the direction of flow and the upper part consists as much as possible of surfaces "sheltered" from the wind. *Sänger* found that a fuselage nose with a ogive shape might be suitable for *Mach* speeds up to *Mach 7*, whereas, the semi-ogive shape to be more suitable for speeds up to *Mach 20.*

TABLE OF CONTENTS

Were it not for *Professor Walter Georgii*, head of the *DFS* and *Eugen Sänger's* boss, the report known as "*Uber einen Rakienantrieb für Fernbomber*" would not have been published. *Walter Georgii* had influence in the aviation community, especially since the *DVL*, where *Eugen Sänger* and *Irene Bredt* worked prior to joining *DFS*, had refused its publication. Shown is the "Table of Contents" page from *Eugen Sänger's* August 1944 proposal: *A Rocket Drive For Long Range Bombers*.

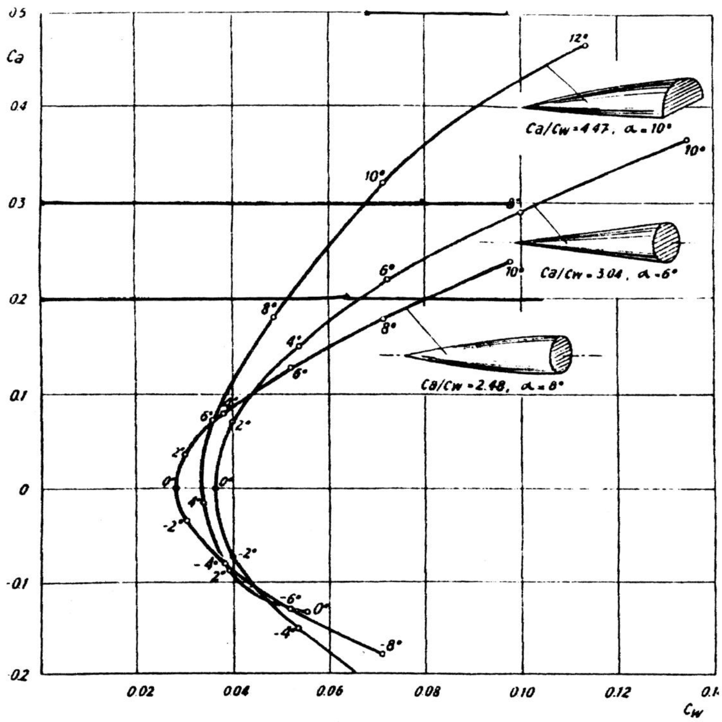

Polar curves of bodies of same thickness and frontal area, placed obliquely to the direction of flow (cone, ogive, and semi-ogive) in a three-dimensional flow. They have the same frontal area and the same height, and their configuration would be suitable for a fuselage nose. *Eugen Sänger* reasoned that the semi-ogive would be best suited for his orbital bomber. This is because the cone and ogive are approximately equivalent as regards lift/drag ratio. The semi-ogive engaged at its flat cross-section, reveals a considerable superiority, having an optimum lift/drag ratio of 4.12.

One of the actual wind tunnel scale models carved out of a metal alloy (1944-1945) for the proposed *Sänger* winged orbital reusable rocket bomber. Featured is its port side.

- They doubted that *Sänger* could deliver a thrust exhaust velocity of 9,900 to 16,500 feet/ second;

- They doubted that *Sänger* could achieve and sustain a combustion chamber pressure of 1,470 pounds per square inch;

- *Sänger* had not satisfactorily dealt with the problems of his orbital bomber's ex-atomospheric 3-axis control or reentry;

- *Sänger's* proposed use of a 1.8 mile long takeoff ramp appeared to have no advantages over that of a vertical lift off;

The Soviet Union never did get around to a design/development program for the *Sänger* orbital bomber. *Tokaev* said that his country had no experience in the field of high altitude rockets, lacked the specialized people, and lacked the research institutes. In addition, Soviet metallurgists had not yet been able to produce high temperature alloys having sufficient s heat-resisting qualities for use as a rocket combustion chamber. Nevertheless, the Soviets – who became intrigued by the idea of a round-the-world space bomber – sought to kidnap *Sänger*. During this period, *Sänger* was living in Paris and working with a trusted group of Germans, including former *DFS* leader *Professor Georgii*. Although they searched for months, *Tokaev* and *Vasilli Stalin* couldn't locate this group. During his search in France, *Vasilli Stalin* seemed to become more interested in the Paris nightlife than in finding *Sänger*.

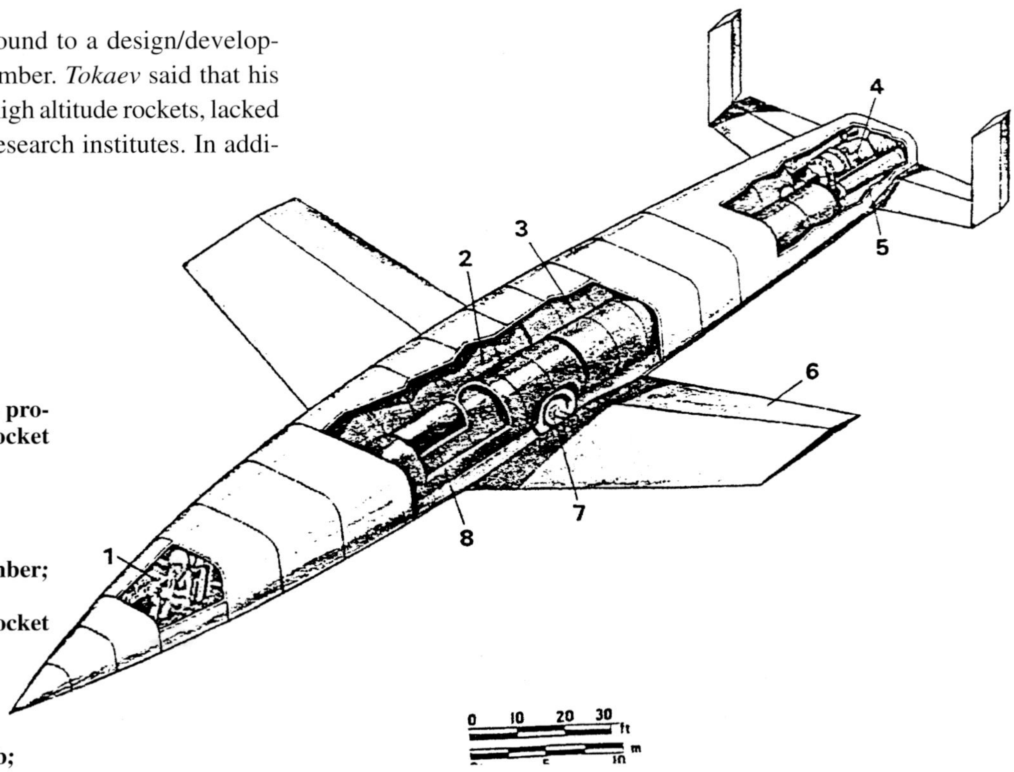

A pen and ink cut away illustration of the proposed *Sänger* winged orbital reusable rocket bomber by *Eugen Sänger*:
01 - **pressurized cockpit cabin;**
02 - **oxidant tanks;**
03 - **fuel tanks;**
04 - **single high-pressure combustion chamber;**
05 - **auxiliary rockets;**
06 - **the aileron control surfaces of the rocket bomber's wedge-shaped wing of *Eugen Sänger's* own design;**
07 - **port side main landing gear shown retracted;**
08 - **a single 5,000 pound free-falling bomb;**

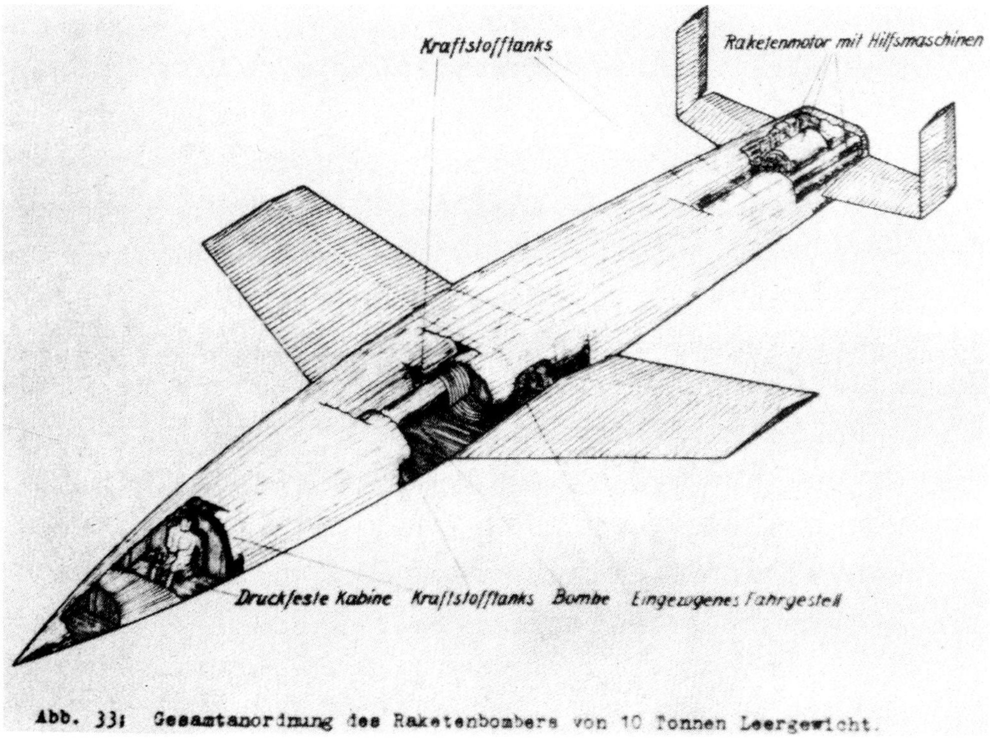

A copy of the pen and ink cut away illustration of *Eugen Sänger's* winged orbital rocket bomber from his secret report of 1944-1945. Notice that in addition to the single main combustion chamber (*raketenmotor*) shown port and starboard are "*hilfsmachinen*" help or assistance machines . It is assumed that these *hilfsmachien* were auxiliary rocket motors, perhaps for maneuvering at 80 miles altitude. Fuel burned, thrust, and duration unknown.
01 - ***Druckfeste Kabine*** **- pressurized cabin;**
02 - ***Kraftstofftanks*** **- fuel tanks;**
03 - ***Bombe*** **- bomb**
04 - ***Eingezogenes Fahrgestell*** **- retracted landing gear;**
05 - ***Kraftstofftanks*** **- fuel tanks;**
06 - ***Raketenmotor mit Hilfsmaschinien*** **- rocket motor with auxiliary engines;**

Tokaev said later that they believed *Sänger's* talented assistant, *Irene Bredt*, who was an outstanding mathematician, also was working with the *Georgii*-led group. "I, personally, looked for her even more assiduously than for *Sänger* himself, but no trace of *Irene Bredt* was ever found," recalled *Tokaev*. Nevertheless, according to *Tokaev*, the search team *was* able to assemble a sufficient number of German scientists and relevant research data in the Soviet Zone of occupation to enable Russian scientists to carry on with rocket developments that had been stalled by Germany's loss of the war. Most German rocket scientists, suspecting a less secure fate under Soviet occupation and attracted by offers of money and research opportunities in the West, had fled to the British, French, or American zones before the end of the war. *Tokaev* came to the conclusion that the realization of *Sänger's* rocket bomber idea was possible in principle, but it was also his considered opinion that the USSR did not possess the means to realize it at that time.

Characteristics of *Sänger's* Rocket Motor

Eugen Sänger actually considered his 100-ton thrust rocket motor a comparatively simple machine. The fuel would go from the fuel

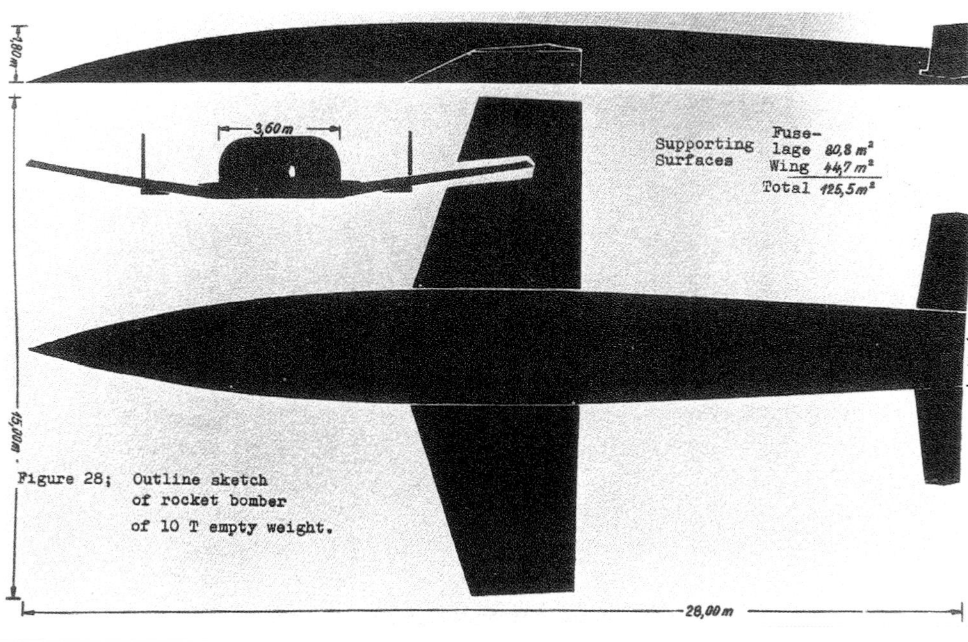

A reverse 3-view image of *Eugen Sänger's* winged reusable orbital bomber featuring its dimension in metrics. Notice that the draftsman has written that the machine's empty weight is 10-ton and that it appears to be powered by two 100-ton rocket engines.

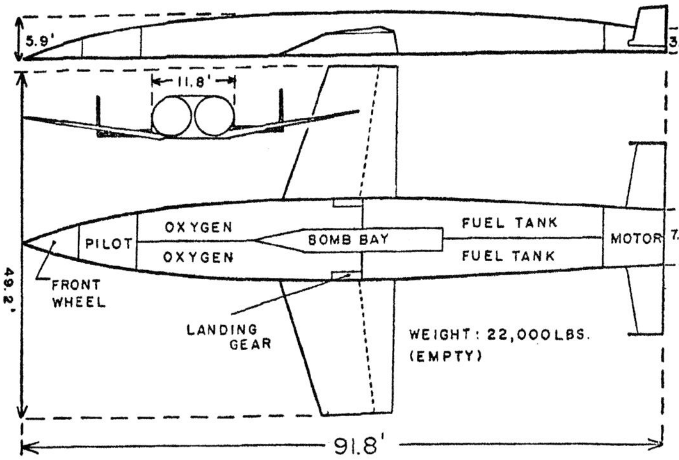

A pen and ink 3-view drawing of *Eugen Sänger's* winged orbital rocket bomber with dimensions in feet and inches. Date unknown. This drawing is included in this book for reference only. One has to be careful about believing in it completely, for example, the draftsperson has written in the lower right hand corner that the empty weight of the *Sänger* machine is 22,000 pounds or 11-ton. This is not correct, because we know that its empty weight would have been 10-ton or 20,000 pounds. In addition, the draftsperson appears to have drawn in two *Sänger* 100-ton thrust rocket engines in its tail. *Eugen Sänger* calculated that his machine would climb to 500,000 feet or 93 miles altitude within 8 minutes with only one 100-ton thrust rocket engine. So why two?

A pen and ink illustration of the cockpit of *Eugen Sänger*'s piloted, winged, sub-orbital, bi-liquid fuel rocket bomber. This cockpit arrangement for the bomber is strictly provisional...the cockpit may have been internal like that shown, or it may have had a raised cockpit. *Sänger's* machine is frequently presented with an internal cockpit, but in fact, it is not known how it would have appeared in the prototype.

tank to the fuel pump, where it was compressed to 150 atmospheres. It would then be fed continuously through valve #5 to the fuel injection head of the combustion chamber. The oxygen came from the thin-walled, uninsulated oxygen tank into the oxidizer pump and was compressed to 150 atmospheres. It was then forced through valve #6 and through the pipe system of the condensers into the oxidizer injection head of the combustion chamber after being warmed to 0 degrees Centigrade. In the combustion chamber, the fuels come together for the first time, mixed, and burned producing exhaust gases at a constant pressure of 100 atmospheres and with a temperature of 4,000 degrees Centigrade. In the heart of the burner chamber, the combusting gasses expand with a very high pressure and form the driving master jet having an exhaust velocity of 3,000 to 4,000 meters/second [9,842 to 13,123 feet/second]. This reaction consumes fuel at the rate of 245 to 327 kilograms/second [540 to 721 pounds/second] and producing a thrust of 100-tons. With a 90-ton fuel supply, the aircraft's rocket motor could be operated at a rate of 100-tons of thrust for 275 to 367 seconds.

About October, 1938, the *Sänger* team fabricated a 1/20 scale model of his concept of a plano- convex supersonic glider aircraft

An artist's rendering of the proposed *Eugen Sänger* winged orbital rocket bomber in space. Notice the stars throughout the illustration. Featured is the machine's port side with a cut-away view of its engine and combustion chamber.

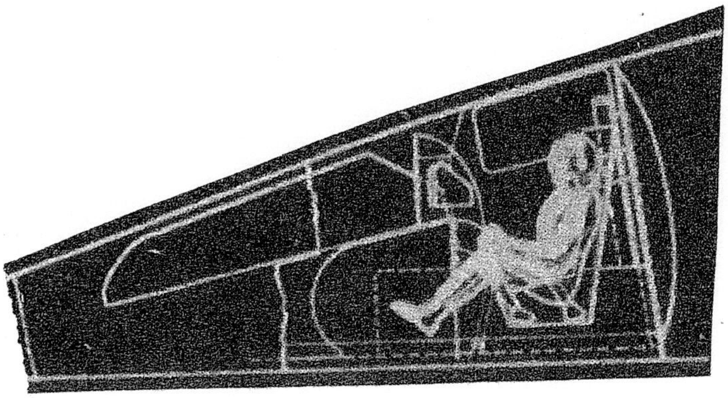

A pen and ink illustration featuring the starboard side view of the cockpit on *Eugen Sänger's* winged reusable sub-orbital, bi-liquid fuel rocket bomber. This illustration is provisional, because no internal drawings of the machine are known to have been drafted.

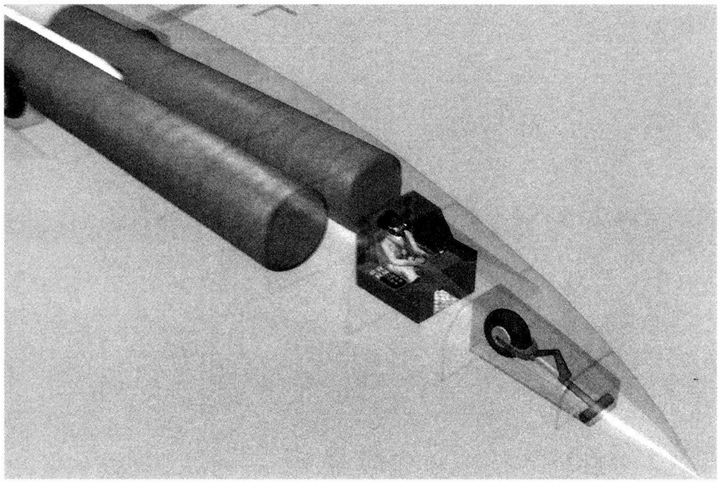

The nose starboard side of *Eugen Sänger's* proposed piloted, winged, reusable, sub-orbital rocket bomber of 1942-1944. Directly aft the single seat pressurized cockpit is the flying machine's liquid oxygen tanks followed by the fuel tanks.

design which would be used for wind tunnel testing. He was now calling his prototype rocket-powered transport flying machine the "*Silbevogel*" or "*Silverbird.*" It's optimal lift- drag ratio assumed a landing velocity of *Mach 0.12*, and was a winged vehicle having a wing mounting angle of 5 degrees incidence. On the basis of initial wind tunnel tests results, *Sänger* applied for a patent for his proposed airframe type with its half ogival-shaped fuselage, bullet-shaped nose, and wedge-shaped wing profile. Due to its dome-shaped body profile and flat bottom, *Sänger's* assistants nicknamed it the "*flat iron.*" Later, the German *Reichspatent* # 411/42 concerning "Gliding bodies for flight velocities above *Mach 5,*" was granted on June 3rd 1942, effective from April 22nd 1939.

To save on-board vehicle fuel weight, *Sänger* proposed accelerating his "*Silverbird*" transport aircraft before lifting off, to a velocity of about 500 meters/second [1,640 feet/second] by means of an auxiliary rocket-driven launching support sled sliding on a straight, horizontal, steel rail 3.0 kilometers [1.86 miles] in length. Thus, it became necessary to obtain knowledge about the amount of dynamic friction between the sliding surface and the upper face of the rail. Considering the very high sliding velocities and the subsequent high negative decelerations required for braking, *Sänger* had to ensure a reliable dynamic floatation of the sliding surfaces by choosing a suitable geometry of the lubricating gap, and then select an optimum lubricant. Because there was no reference data

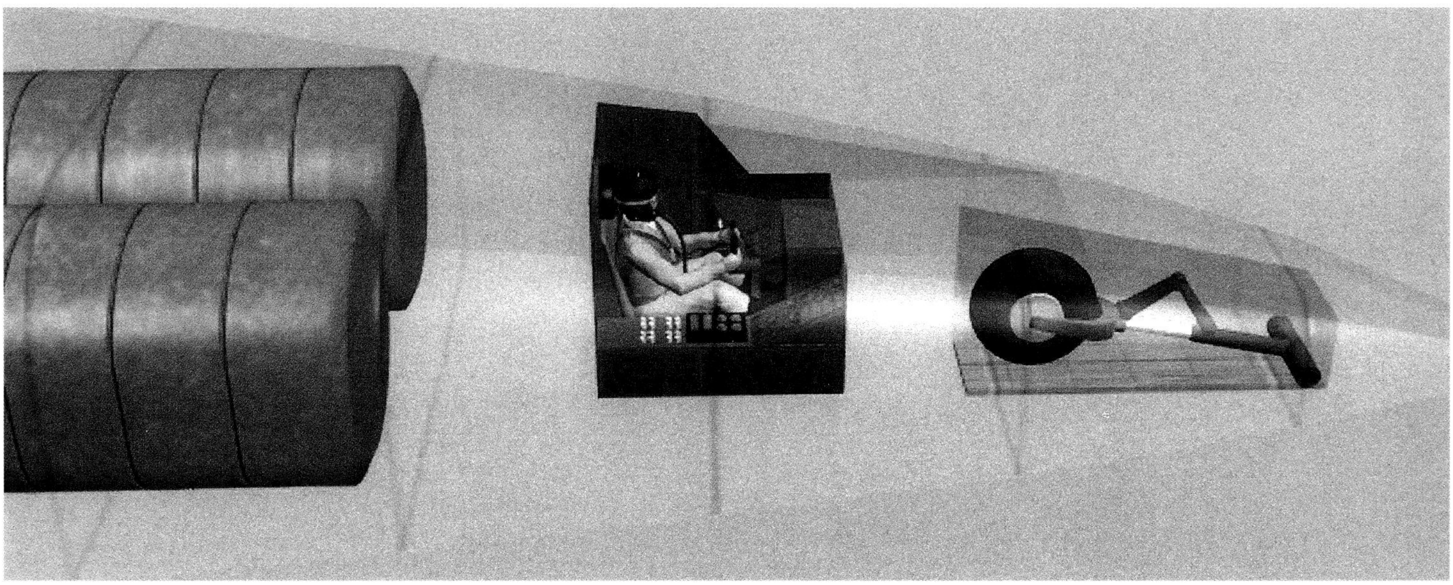

The general position of the pilot's cockpit in *Eugen Sänger's* proposed piloted, winged, sub-orbital, bi-liquid fuel rocket bomber. Featured is the flying machine's starboard nose side. Digital image by *Mario Merino*.

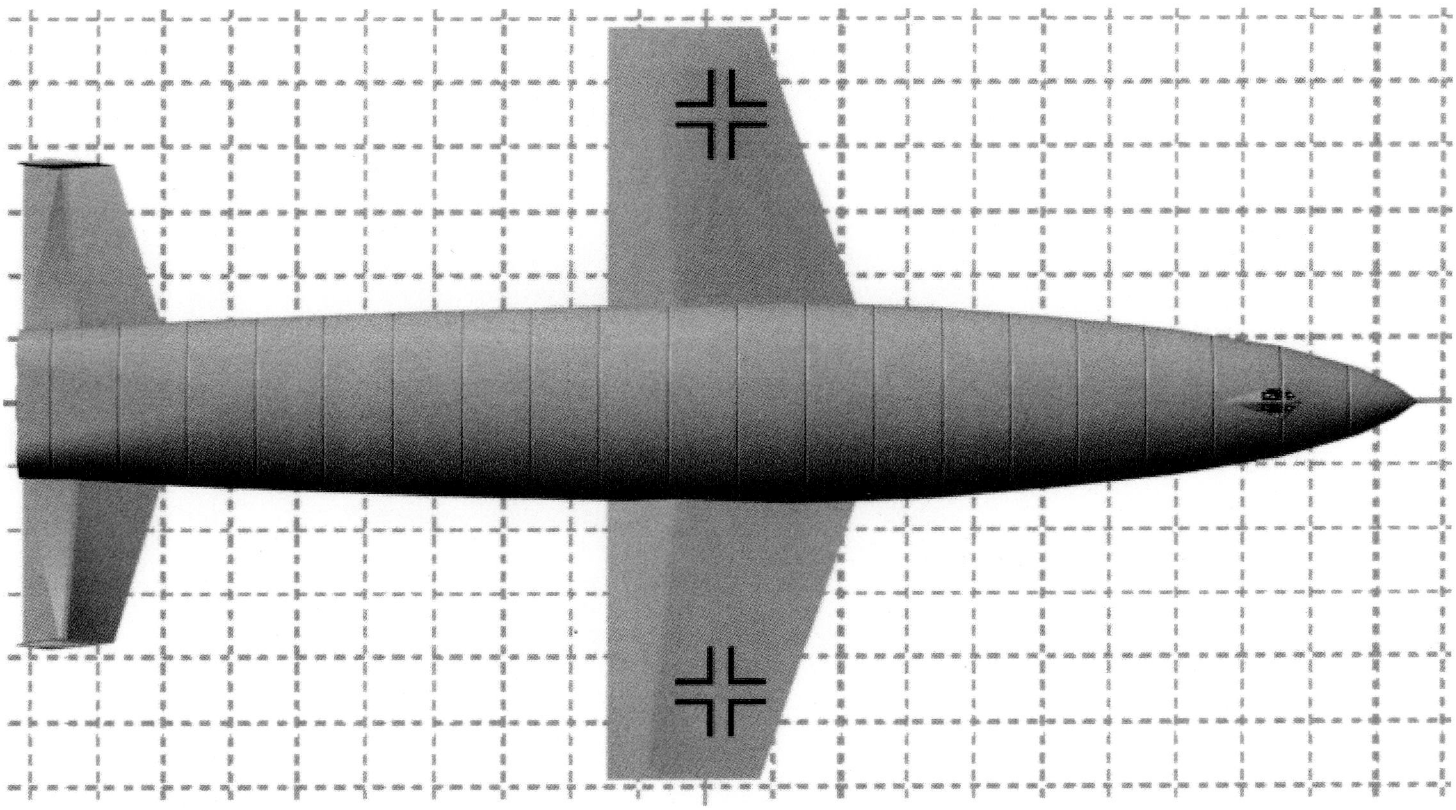

Overall planform (top) of *Eugen Sänger's* proposed piloted, winged, sub-orbital bi-liquid fuel rocket bomber. Digital image by *Mario Merino*.

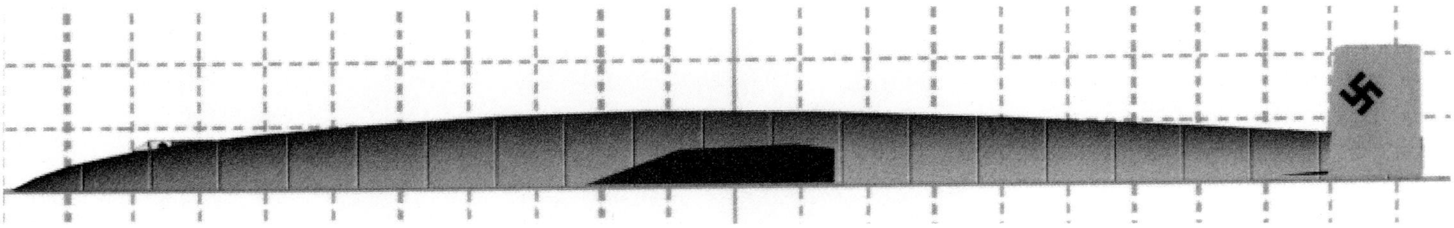

Outline of *Eugen Sänger's* proposed winged reusable orbital rocket bomber as seen from its port side. Digital image by *Mario Merino*.

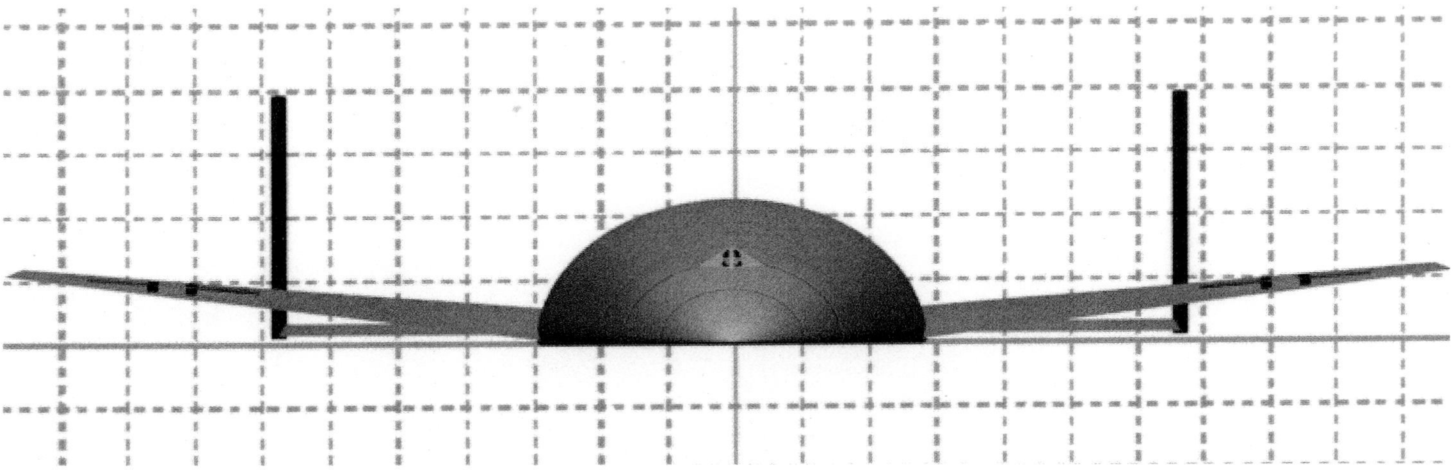

Outline of *Eugen Sänger's* proposed piloted, winged, reusable sub-orbital rocket bomber as seen head on (nose). Digital image by *Mario Merino*.

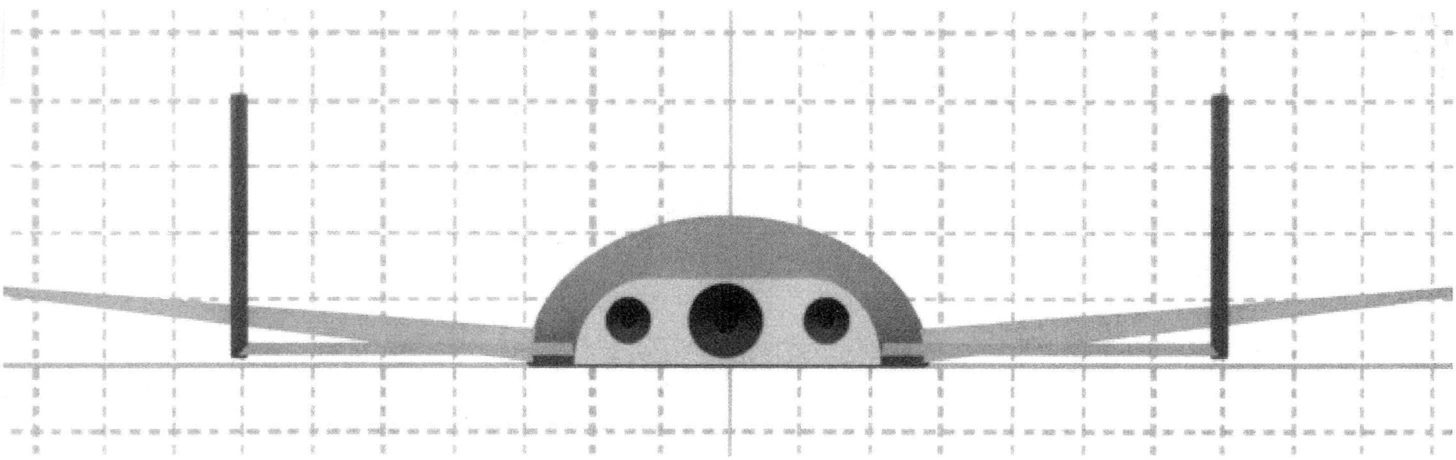

Outline of *Eugen Sänger's* proposed piloted, winged, sub-orbital rocket bomber as seen from the rear. Notice that this outline includes a single combustion chamber with a port and starboard side maneuvering solid rocket thrusters. Digital image by *Mario Merino*.

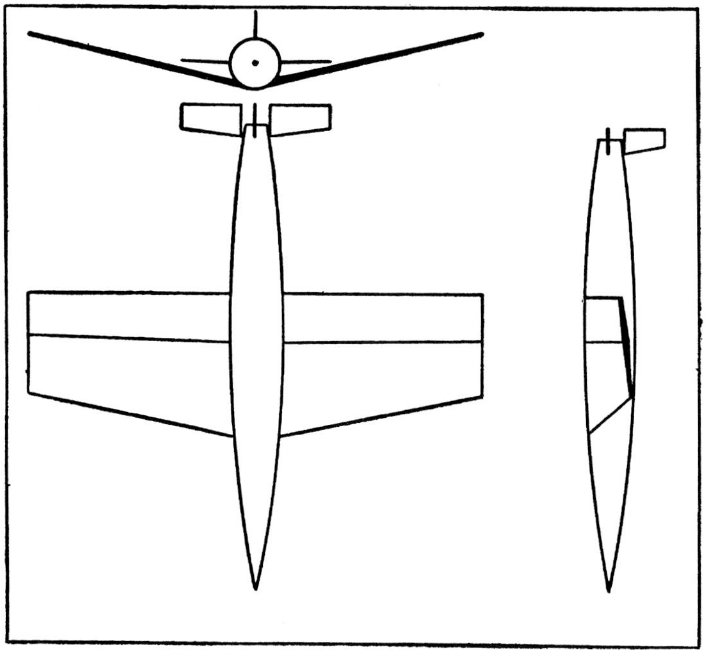

in the literature at the time regarding such extreme operating conditions, some experts were stating that it would be impossible to control the frictional heat generated and, therefore, the concept of *Sänger's* idea of using a horizontal launching ramp, versus a vertical lift-off option, became questionable. *Sänger* asked his assistant *Irene Bredt* to investigate and collect proven values for dynamic friction and lubricating-procedures from qualified experts. However, research had not yet been directed at experiments with sliding velocities of several hundred meters per second. The highest sliding velocity data then available came from research professors *Fottinger* and *Vogelpohl* at the *Institut für Technische Strömungsforschung*-Berlin (The Institute for Technical Flow Research at Berlin) and were limited to a fraction of the required values proposed by *Sänger*. It was *Sänger* himself who came up with a

Left: A view of *Eugen Sänger's* wind tunnel winged rocket bomber scale model as seen from its bottom port side. Notice that the fuselage underside is flat. It was designed this way, according to *Sänger*, to give the orbital rocket bomber a more effective skipping performance between dense and less dense atmospheric layers.

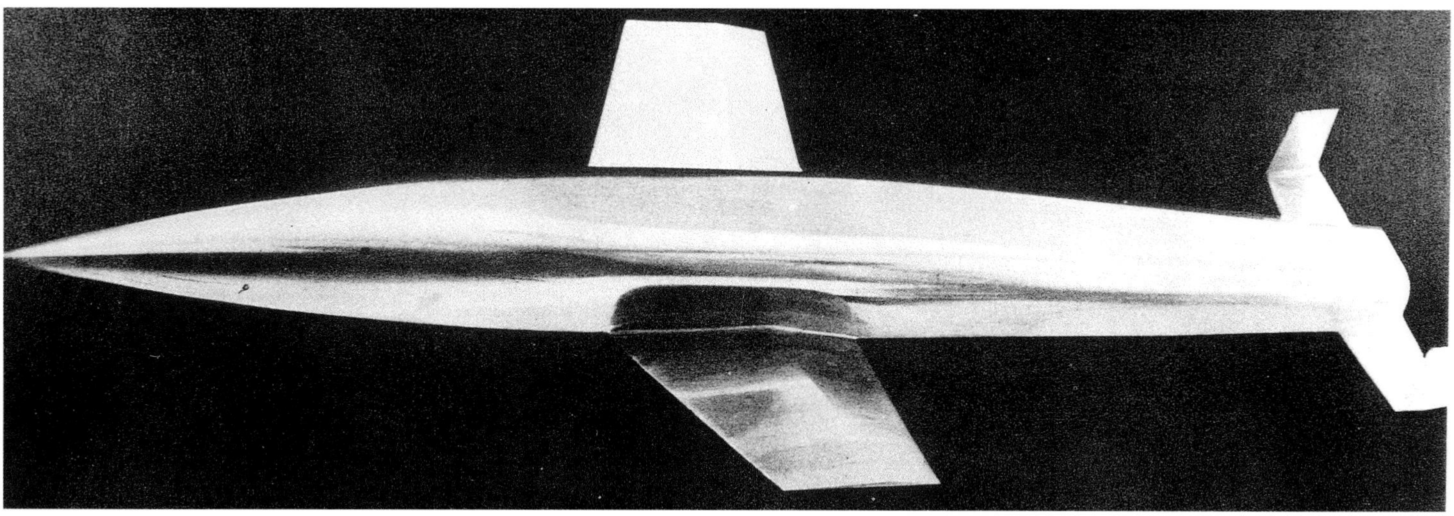

Wind tunnel scale model of *Eugen Sänger's* winged reusable sub-orbital rocket bomber as seen from its nose port side.

Wind tunnel scale model of *Eugen Sänger's* winged sub-orbital rocket bomber as seen from its top port side.

plan. He suggested a series of high velocity sliding- friction experiments, which began on June 2nd 1939, and ultimately demonstrated that construction and use of precision sliding surfaces for velocities up to 500 meters/second [1,640 feet/second] was possible on a carefully finished and lubricated rail.

Sänger and his assistants tested the basic feasibility of this type of sliding launch at high speeds. *R. Schmid* made tests on the sliding properties of projectiles along curved metal walls lubricated with *Vaseline*. The experimental arrangement consisted of a drawn-steel tube of 10 mm [about 1/2 inch] inner diameter, which was slit along its length axis to produce a U-shaped groove. This channel then was bent into a closed circle 8 meters [26.2 feet] in radius so that the groove surface of the channel was toward the outside. Therefore any projectile sliding in the channel would be kept there by centrifugal force. Then, through a branch tube which was mounted tangent to this circle, a stainless steel projectile was fired from a German *'Mauser'* rifle into the circular path. The bullet ran through the circular path several times until its muzzle velocity dissipated and it came to rest. After the test, the bullet was found lying completely unharmed in the channel. The thin copper plating of the projectile was rubbed off at the bullet's maximum diameter. The steel surface also showed isolated scratches, apparently caused by filings and burrs along the slide-path. *Sänger* said the bullet was still warm to the touch immediately after it was located. Also it was slightly compressed in cross-section, because the initial spinning of the bullet (caused by the riflings in the barrel of the gun) stopped

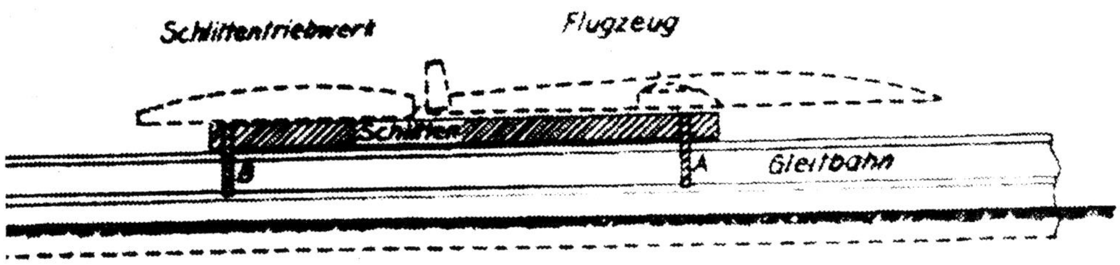

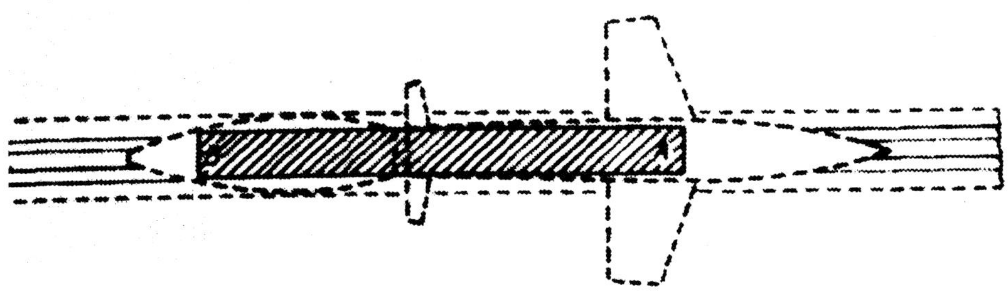

A pen and ink drawing showing the general arrangement of *Eugen Sänger's* piloted, winged, reusable, orbital rocket bomber's launch monorail with its *von Braun's A-4 (V2)* rocket engine equipped propulsive captive sled pushing the bomber down its 1.8 mile long monorail launch track.

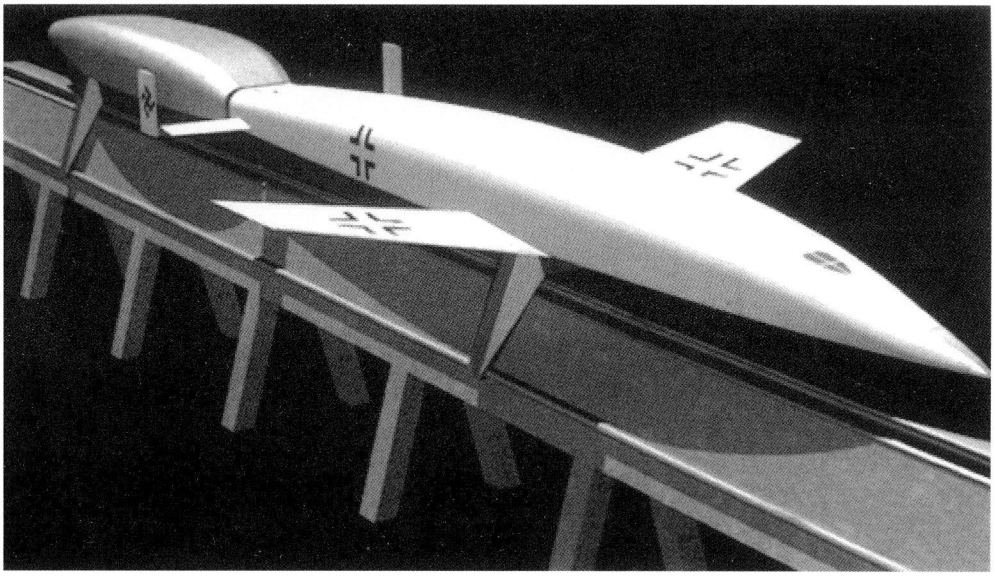

The *Eugen Sänger* winged orbital rocket bomber atop its 1.86 mile long monorail launch with its *von Braun A-4 (V2)* rocket engine equipped captive sled seen from their starboard side. Digital image by *Josha Hildwine*.

after a short time and then the heat of friction caused by centrifugal force of the bullet traveling around this channel, which was initially 100 kilograms [220 pounds], caused the lead core in the bullet to soften and flow. The results of this elaborate test showed that the bullet experienced a range of velocities beginning at 800 meters/second [2,625 feet/second or 1,790 miles/hour] down to zero without being harmed. *Sänger* believed this test proved the feasibility of constructing a launching rail to support a sled that would reach velocities of 500 meters/second [1,640 feet/second or 1,075 miles/hour], provided the slide-rails on this monorail track could be aligned with precision and kept well-lubricated.

A few weeks after *Sänger's* successful sliding-friction experiments on the heath near Trauen/Fassberg, *World War II* started on September 1st 1939. This now meant that *Sänger's* small staff at the rocket-research establishment on the quiet heath of Lüneberg, could expect substantial changes in the freedom allowed in their research work. *Sänger's* group had been living in a sort of dream world, spending all their working hours, as well as free time, in the remote Oertze Valley location. This environment was rather isolated from other people and the general political events of *Nazi* Germany at that time. *Sänger* and his group created their own separate virtual 'reality.' All this tranquility on the Lüneberg Heide changed virtually overnight with the onset of the war, however. According to *Sänger*, there now were priority schedules, mobilization, calling up orders, and soon thereafter, an increasing general shortage of strategic materials. The continuation of *Sänger's* research work and

A view of *Eugen Sänger's* winged orbit rocket bomber, its captive propulsive sled, its monorail built above a marsh and seen from its port side. Digital image by *Mario Merino*.

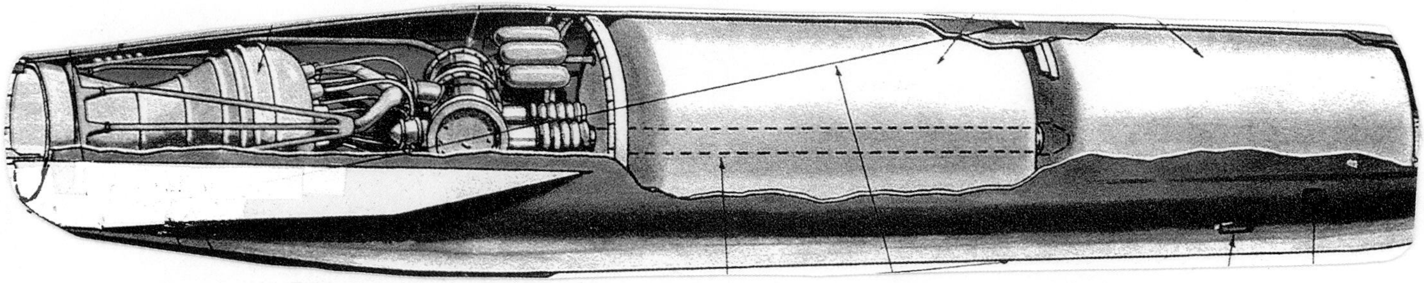

A pen and ink cut away starboard side view of *Sänger's* proposed *V2* rocket engine equipped propulsive sled shown with a single *V2* rocket engine. Could a single *A-4 (V2)* rocket engine push the 100-ton *Eugen Sänger* winged reusable orbital rocket bomber up to a speed of 1,113 mile/hour within 1.86 miles atop its monorail launch rail? *Al Bowers* of NASA-Dryden Research Center believes that it would have taken as many as 30 *A-4 (V2)* bi-fuel rocket engines.

Remember that our targets are: 1,632 feet per second (1,113 mph) within 11,616 feet (2.2 miles). We have to assume that *Sänger's* propulsive sled was a "full up" *A-4 (V2)* rocket, which had an average weight of 5,562 pounds and an 11 second burn time. We have to assume that it would carry only enough fuel for its 11 second burn. So each *A-4 (V2)* unit adds that much to the total sled (assumed weight is zero) weight.

Number of *A-4 (V2)* rocket units	Speed (ft)	Distance (ft) after 11 seconds	Thrust (lbs) after 11 seconds
1	68,500	118.0	649.4
2	137,000	229.8	1,264.1
6	411,000	624.8	3,430.9
12	822,000	1,092.5	6,003.3
18	1,233,000	1,455.2	8,003.6
24	1,644,000	1,746.1	9,603.5

Given aerodynamic drag, it would probably have taken even more than 24 *A-4 (V2)* rocket units (perhaps as many as 30), and their associated fuel tankage to achieve 1,113 mph speed in only 1.86 miles.

experiments at the "*Flugzeugprufstelle,*" the camouflage name for the rocket testing stands at the *Hermann Göring Institute-Trauen/Fassberg*, appeared to be in danger because these projects were deemed unsuitable for immediate military use. For *Sänger* to continue on with his research work for a civilian transport flying machine, a proven 'military application' had to be created to justify current work. Related 'auxiliary projects' had to be conceived – projects that promised advantageous military application within a reasonably short time. One such research project *Sänger* bid for concerned a fighter airplane with ramjet propulsion. This project took on a greater importance and priority as the war continued.

The interim results of *Eugen Sänger*'s "*Silverbird*" research project had just been summarized in a secret report – with about 125 copies printed – having the title "*Rocket Space Plane.*" *Sänger's* heretofore civilian research project now had to be dressed in 'new clothes' if it was to survive. Thus, *Sänger* renamed his research, calling it "investigations into a rocket bomber project" or *Raketenbomber*. In addition, *Sänger* rewrote his earlier summarized

A close up view of the starboard side of *A-4 (V2)* propulsive unit being examined by U.S. Army officers. Nordhausen-Thüringen, May 1945.

A flight ready *A-4 (V2)* under armed guard by an American soldier and featuring its port side propulsive unit. May 1945.

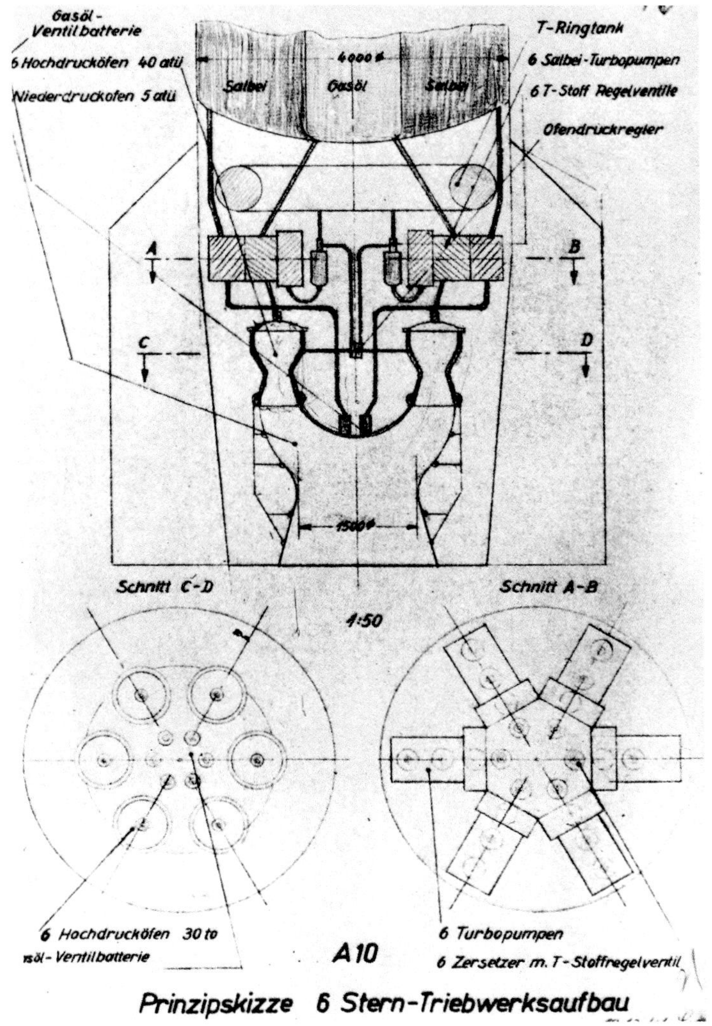

A pen and ink drawing featuring six *A-4 (V2)* rocket engines grouped together for *von Braun's* proposed *A-10* booster stage rocket for lifting the piloted *A-9* into sub orbit. On the other hand, *Eugen Sänger's* piloted, winged, reusable, sub-orbital bi-fuel liquid fuel rocket bomber might have required as many as 30 *A-4 (V2)* rocket engines grouped together in a single propulsive sled.

report for a rocket-powered vehicle, adding several new chapters covering the trajectories of bombs, impact ballistics, and offensive measures. Thus, *Sänger's* former rocket-powered civilian space transport airplane project now evolved into an Earth-orbiting, single-stage, rocket-powered intercontinental bombing machine with a launch weight of 100-tons – with 86- to 90-tons of propellant (depending on tonnage of bombs), 9-tons of airframe and engine, and a 1- to 5- ton payload of bombs. It would be propelled by a rocket engine using highly efficient fuels with liquid oxygen used as an oxidizer in a combustion chamber at a pressure of 100 atmospheres and creating 100 tons of thrust. The maximum flight performance

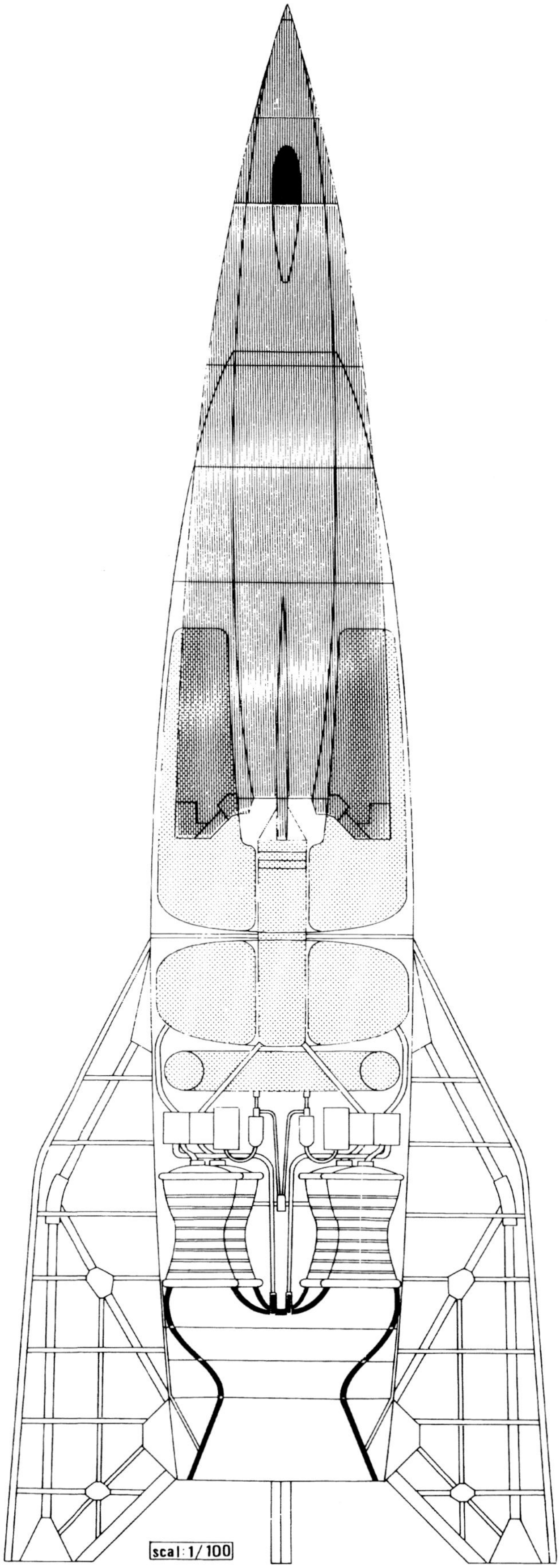

Left: A pen and ink illustration featuring *Wernher von Braun's A-10* booster with a piloted *A-9* 2nd stage rocket missile mounted on top of the booster. Courtesy: *J. Miranda & P. Mercado, Secret Wonder Weapons of the Third Reich,* Schiffer Publishing, 1996.

of this rocket plane, based on the assumption of semi-ballistic flight trajectories and of a specific impulse (rocket ignition) lasting 6 1/2 minutes was calculated as:

- Flight velocities at the end of the power flight phase of about 2,625 feet [800 meters] per second, corresponding to the necessary thrust for reaching orbital velocity with a 1-ton [2,000 pound] payload;
- Flight altitudes in the ballistic section of the flight path up to 186 miles in height [300 kilometers];
- Loading (payload) capacities for a transport to the antipodal point of the Earth 12,428 miles distant [20,000 kilometers] of up to 8-tons [16,000 pounds];
- Ability to achieve flight distances up to a single Earth orbit with a payload of 4-tons, or up to 2 1/2 orbits with a 1-ton [2,000 pounds] payload.

A pen and ink illustration featuring the rocket engine arrangement in *Wernher von Braun's A-10* booster statge. Courtesy: *J. Miranda & P. Mercado, Secret Weapons of the Third Reich,* Schiffer Publishing, 1996.

These flight performances of the former civilian rocket-powered space plane – turned rocket-powered intercontinental bomber – became possible by the application of a semi-ballistic flight technique proposed by *Eugen Sänger*. He later called it *"Rikoschettier"* or "skipping flight", where the aircraft ricocheted from the dense layers of the atmosphere like a stone flung at a flat angle and 'skipped' across the surface of a pond. In this way, gliding flight paths could be obtained which would be several times the range obtainable with only a conventional aerodynamic ascent and descent. *Sänger* then went on to rewrite the findings of his earlier research and he conceived a new, more militarily-attractive title: *"On a Rocket Propulsion Engine for Long-Distance Bombers"* or *Ubereinen Raketenantrieb für Fernbomber*. The nearly 900-page draft was completed and submitted for approval to the *RLM* on December 3rd 1941.

A ground level view of *Eugen Sänger's* proposed winged orbital rocket bomber being propelled by its several *V2* bi-fuel liquid rocket engine powered sled as it begins its roll down its 1.86 mile long horizontal monorail launch. Digital image by *Josha Hildwine*.

Proposed 12 Steps to the Rocket Bomber's Development

Sänger said in order to make his rocket bomber a reality, a systematic process for development would be required. He did not give a timetable for the development of his rocket bomber, but he did propose that the following sequence of 12 phases would be required in order to make the rocket bomber a reality:

***Phase 1* – Development of the Combustion Chamber and Jet Nozzle**

Sänger believed that the main problem confronting development of the rocket bomber was the configuration of the engine. He was concerned about the introduction of various types of solid, liquid, or already vaporized fuels, and the combustion-supporting oxidizer material (perhaps super-cold) into a combustion chamber through injection nozzles. Then there was his concern involving the rapid distribution, mixing, heating, and ignition of the fuel and oxidizer mixture to give complete combustion at a constant high pressure. *Sänger* stated in his report that his rocket bomber would require fuel being fed at a constant high pressure into a combustion chamber operating at very high temperatures in order for the gas to expand rapidly into the form of a combustible vapor. Then this vapor would be ignited and converted to a stream of exhaust gas with as high a velocity and as low a temperature as possible. *Sänger* realized that sustained high pressures and temperatures in the combustion chamber could present insurmountable engineering problems. Research was only beginning in Germany at the time on high-strength alloys capable of resisting extremely high temperatures and pressures simultaneously. *Sänger* assured the readers of his proposal that the solution to these problems rested in the choice of materials, methods of cooling, and constructional arrangement for the powerplant, and that these considerations should be the main point of study in the bomber's first phase of development. *Sänger* believed that this study would also answer other questions, such as:

A ground level view of how *Eugen Sänger's* captive propulsive sled with its 30 *von Bruan A-4* (*V2*) rocket bi-fuel liquid fuel engines might have appeared sitting atop its monorail launch ramp. Digital image by *Mario Merino*.

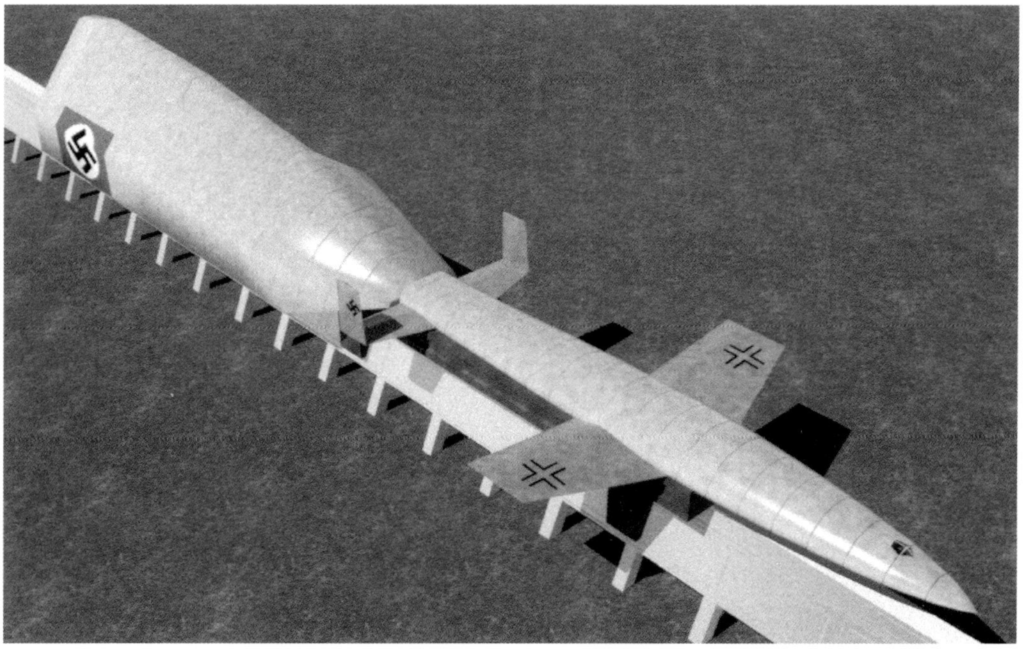

An overhead view of *Eugen Sänger's* winged reusable orbital rocket bomber and its 30 *A-4 (V2) von Braun* bi-fuel liquid fuel rocket engines in a captive propulsive sled. Digital image by *Mario Merino*.

- The most suitable shape and relative size of the rocket bomber's combustion chamber;
- The choice of the most suitable flame pressures and methods for measurement of exhaust stream temperature and velocity;
- Arrangements for rapid heating, mixing and vaporizing of fuel;
- Optical and acoustic phenomena;
- The mixing of the exhaust vapor with the surrounding air behind the motor;
- Dissociation and detonation problems;

***Phase 2* – Development of Special Fuels for Rocket Motors**

Sänger emphasized to his readers that rocket engines require very different fuels than those commonly used in ordinary aircraft engines. He wrote that the principal concern was the available *energy content* per unit mass of the combustible of the fuel and oxidizer mixture and not the mere thermal value of the fuel alone. He theorized that a combustible material having a lower heat of combustion than typical hydrocarbon fuels, but consuming much less oxygen during combustion, could develop a far superior specific en-

Eugen Sänger's* piloted, winged, reusable sub-orbital rocket bomber approaching 1,113 miles/hour forward speed as it nears the end of its 1.86 mile long launch. Notice how the trees in the background are a blur as the speeding orbital bomber roars past on its monorail at nearly twice the speed of sound. Digital image by *Mario Merino

The *Sänger* winged reusable sub-orbital rocket bomber seen moments after leaving its 1.86 mile long monorail catapult/launch heading east after reaching 500,000 feet [94 1/2 miles] altitude. This digital image features the *Sänger* machine powered by three bi-fuel liquid rocket engines...one large and two smaller units. The number of propulsive units had not been settled upon by 1944, but *Eugen Sänger* claimed that his 100-ton weight bomber with its single 100-ton thrust engine could reach over 100 miles altitude between 4 and 8 minutes. Redoing *Sänger's* math today, based on 1940s rocket technology, the rocket bomber would reach its cruising altitude initially at about 78 to 80 miles altitude in a little over 7 minutes....pretty good performance even by standards of 2001. Digital image by *Josha Hildwine*.

ergy output. In addition to the energy output of the mixture, *Sänger* said, other combustion characteristics, such as ease of ignition; the rate of combustion; any tendency to detonate or explode, rather than burn evenly; the degree of fuel mixture dissociation; the state of aggregation of the combustion products; the ease of supporting the continuous reaction of the fuel-oxidizer mixture; and the resulting reaction temperatures were just as important. He believed that thought also must be given to non-combustion properties such as fuel production and transportation complexities and cost; the ease and safety of fuel storage in tanks on the rocket bomber for indefinite periods under widely varying ambient temperature and humidity conditions; fuel density; risk of explosion from static discharge; and so on. Nevertheless, *Sänger* suggested in his proposal that the principal focus of effort in Stage 2 would be a broad investigation of potentially suitable propellants and he expected the results of this research would lead to development of new and more suitable rocket fuels.

Phase 3 – Development of Auxiliary Engines for the Rocket Bomber

Sänger wrote that a typical rocket motor requires important auxiliary components, such as specialized pumps used for feeding fuel, oxidizer and coolant to the combustion chamber. He mentioned that development of such unique components presented quite daunting engineering challenges. For example, several pumps required a horsepower rating of 50 to100 horsepower per ton of thrust in order to develop sufficient pressures to drive the fuel and oxidizer compounds into the combustion chamber. In addition, the very nature of the material to be conveyed – such as cryogenic liquid oxygen, metallic suspensions, and liquid metals – required very special handling. Such compounds presented vastly different handling qualities and had to be accurately proportioned as well as fed through

***Eugen Sänger's* piloted, winged, reusable, sub-orbital rocket bomber leveling off as it approaches the lower limit of its initial operational altitude of 78 to 80 miles above the earth. Digital image by *Josha Hildwine*.**

A general planform view of the *Sänger* piloted, winged, reusable, sub-orbital rocket bomber as seen from 78 to 80 miles altitude and featuring its starboard side. Digital image by *Mario Merino*.

piping to combustion chambers at very high pressures. *Sänger* also noted another complication had to be considered before studies could get under way – that of using the strongest materials, while designing and manufacturing the machinery under the extreme requirements of light weight. This included machinery for the feeding of fuel and coolant, combustion ignition systems, and new types of pressure and flow regulators.

Phase 4 – Development of a Full-Size Prototype Operational Rocket Motor

Even if the first three phases of development could be successfully completed, *Sänger* wrote, a major accomplishment would be to take all the components of the powerplant and put them together in a ready-to-fly rocket engine. Engineers had to consider the interplay of all components used in a full-sized operational rocket engine. Full-scale operation on a test bench was essential to see the interaction between the combustion chamber, fuel pumps, and other auxiliary components so that suitable adjustments could be made in order to achieve the best performance of exhaust speed, reliability of performance, as well as maximum component strength and minimal weight. *Sänger* considered the bench test phase extremely important because such tests can closely mimic conditions encountered during flight. This stood in contrast to testing procedures used for ordinary aero-engines, where actual flying conditions can be imitated only with difficulty in ground-level testing and not even completely by actual flight tests at altitude. This was mainly due to the fact that the rocket motor accomplishes the combustion, creation of a propulsive exhaust stream, and propellant cooling using only its own fuels and without the use of surrounding atmosphere. One of the few places where the rocket motor came in contact with the surrounding air was at the mouth of the combustion chamber. There the combustion gas expanded rapidly and could be affected by the air behind the aircraft.

Aside from the purely developmental and test studies on the model, a whole series of physical tests on the rocket motor could be carried out most advantageously during Phase 4:

The pilot of *Eugen Sänger's* piloted, winged, orbital rocket bomber would have been the first person to touch the edge of space and view the sun and the stars initially from about 80 to 100+ miles altitude. Digital image by *Mario Merino*.

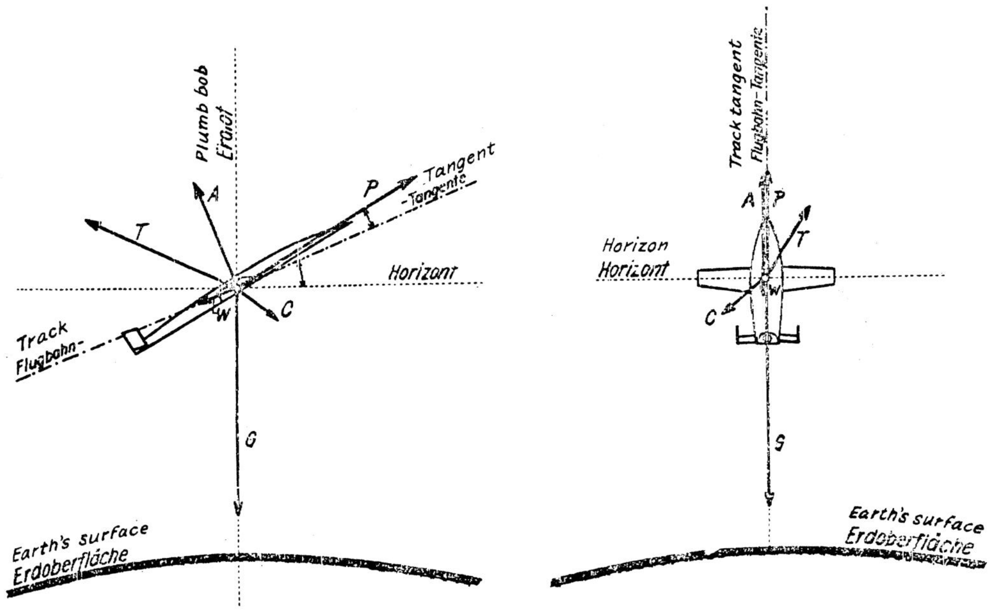

Abb. 59: Äussere Kräfte am Raketenbomber beim Aufstieg.

A pen and ink drawing from the *Sänger-Bredt* document of 1944 referring to the external forces experienced by the winged orbital rocket bomber during its ascent to 500,000 feet or 93 miles altitude. The arrows indicate the force and direction of forces exerted on the orbital rocket bomber as its climbs to the edge of space.
a - **lift factor due to the shape and size of its wings;**
c - **velocity and direction of the orbital bomber if takeoff is aborted;**
g - **gravity;**
t - **atmospheric drag;**
p - **engine thrust;**

- More careful investigation of atomizing the rocket fuel;
- Actual behavior of the pressure and temperature buildup in the combustion chamber;
- Temperature distribution in the combustion and exhaust gases;
- Actual streaming velocities of the combustion gases;
- Condition of the heat transfer from combustion gases to the walls of the chamber, and from there to the surrounding coils carrying coolant;
- The contours of exhaust gases and their sound emission patterns;

Sänger emphasized that one could consider installing a complex rocket engine in an aircraft only after its operational reliability had been proven in static testing on the ground. The development of the rocket bomber airframe would have to parallel that of its rocket motor, so that both could be ready for flight at the same time.

Phase 5 – Wind Tunnel Testing of the Airframe and Aerial Tow Tests

Sänger wrote that the external shape of his rocket bomber had been determined by the requirements of supersonic flight and the fact that the vehicle's propulsion system would be located in the tail of the aircraft. Thus a complete series of wind tunnel tests would be required before proceeding with detailed design drawings for production aircraft. The rocket bomber would depend on gliding-flight at high *Mach* numbers in order to carry out its bombing operations on targets in another hemisphere. At these projected speeds, the

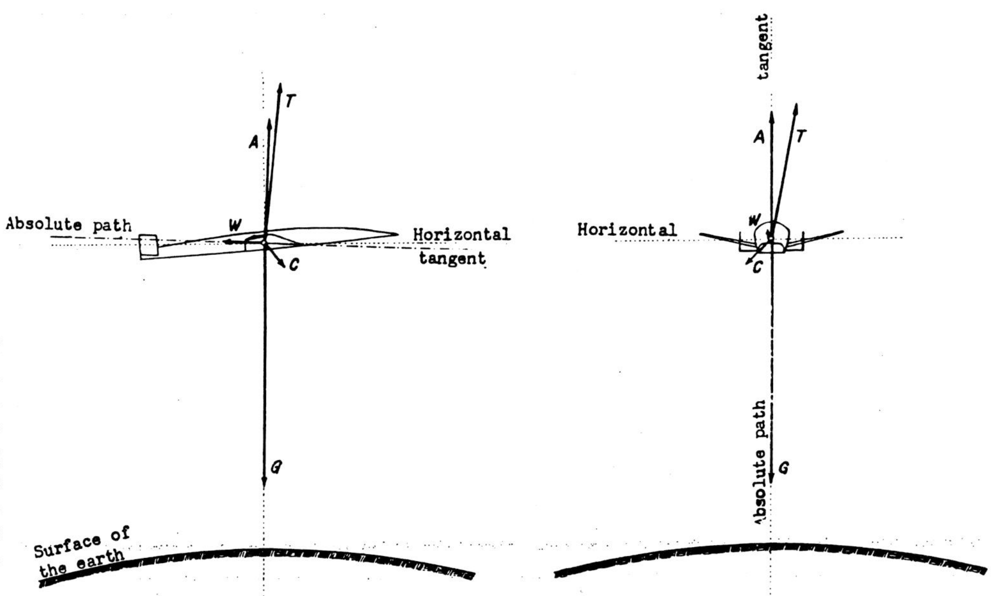

A pen and ink drawing from the *Sänger-Bredt* document of 1944 showing the directional phase before dropping its bomb. The winged orbital rocket bomber pilot was to have determined his position and altitude by fixing on a convenient star in order to release the 5,000 pound radioactive sand wrapped bomb (around Indianapolis, Indiana) so that it would glide down and explode 3,000 feet altitude over New York City.

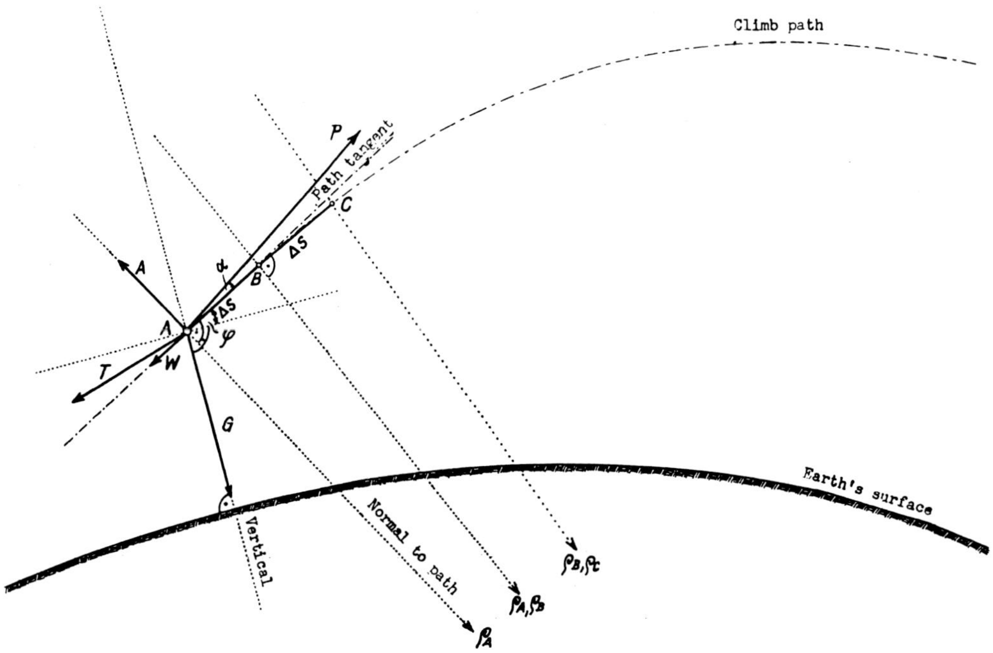

This diagram from *Eugen Sänger's* report features mathematically predicted the path of his winged orbital rocket-powered bomber were it to be flown on its way to a target on the North American continent.

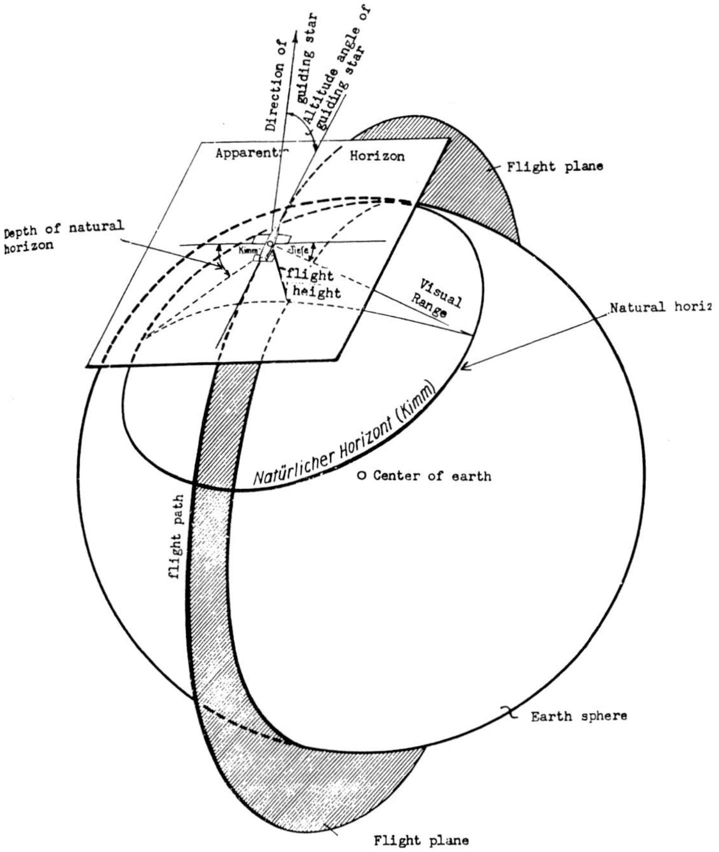

This diagram from *Eugen Sänger's* report features how the pilot of his winged, sub-orbital, bi-fuel liquid fuel rocket-powered bomber would have determined his position in space as well as altitude from earth in order to release his bomb at the exact moment. For example, if the target was New York City and the *Sänger* pilot was approaching from the west at 14,000 mile/hour, then the bomb would have to be released over Indianapolis, Indiana.

aerodynamic forces acting on the vehicle (when still in the Earth's atmosphere) would be proportional to the angle of attack and the square of its velocity. *Sänger* believed that models of the rocket bomber had to be tested in wind tunnels in the speed range of *Mach 0.8* to *Mach 4.0.* He also needed to have data on a range of velocities below the speed of sound, which was important for control of the aircraft during the landing process. So using models of "reasonable size," *Sänger* planned to carry out measurements on the aerodynamic forces – especially the maximum lift of the rocket bomber and the use of various aerodynamic aids for flight control in each specific mission profile. Such tunnel tests would have to be performed to evaluate the stability, vulnerability to vibration, and maneuverability of the machine. In the supersonic flight domain, aerodynamic forces, distribution of air pressure, stability, manoeuverability, vibration conditions, air temperatures at the stagnation points, and heat transfer to the aircraft all are of interest and concern. A whole series of these tests on very small models can be carried out effectively in a supersonic wind tunnel, but there were no such supersonic tunnels in Germany at the time.

For tests with larger models, *Sänger* suggested the use of a monorail takeoff track. He theorized that models could be boosted up to any desired high speed with the aid of rocket engines and then the results studied. *Sänger* said that tests could be carried out at reasonable cost on a 15 to 30 kilometer [9 1/3 to 18 1/2 mile] long track with models scaled in size from 1:10 even up to 1:1 and at full scale speeds of 800 to 5,000 kilometer/hour [497 to 3,107 miles/hour] so that the *Mach* number and *Reynolds* number were exactly duplicated. These tests would be expected to give results on the general behavior of the rocket bomber in the neighborhood of the velocity of sound between 800 and 1,800 kilometers/hour [497 to 1,119 miles/hour] testing various shapes, angles of attack, arrange-

A close up view of a *Sänger* winged reusable orbital rocket bomber moments after releasing its 5,000 pound radiation-spreading bomb. The fuselage's underside was flat enabling the flying machine to better skip along the upper atmosphere like stone skipping over water. Notice the curvature of the earth beneath the orbital bomber. *Al Bowers* of NASA's Dryden Space Facilities estimated that if the *Sänger* bomber was cruising at just under 14,000 miles/hour, its pilot would have to release the bomb over Indianapolis, Indiana, not, as has been suggested in popular post-war literature, over the state line dividing Utah and Nevada. Digital image by *Josha Hildwine*.

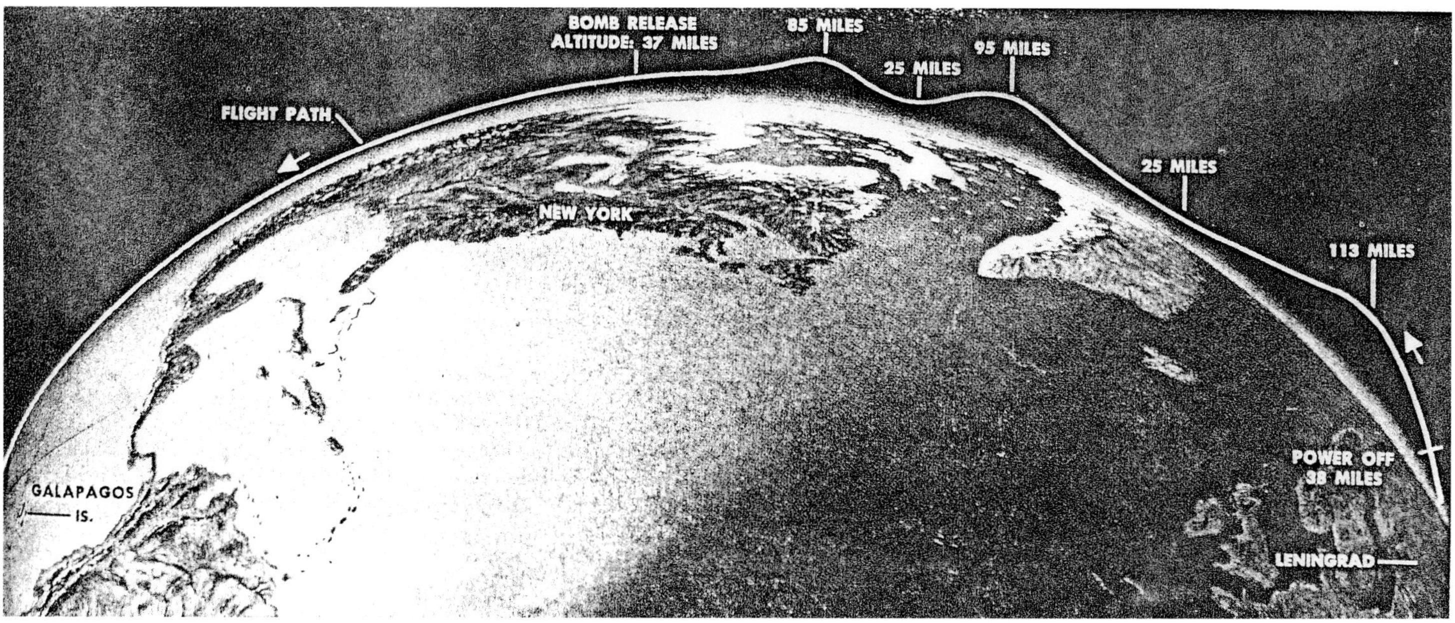

This pen and ink illustration comes of a report from the former USSR and features how the Soviets would have used the *Sänger* bomber concept to bomb America from space after a launching in the USSR.

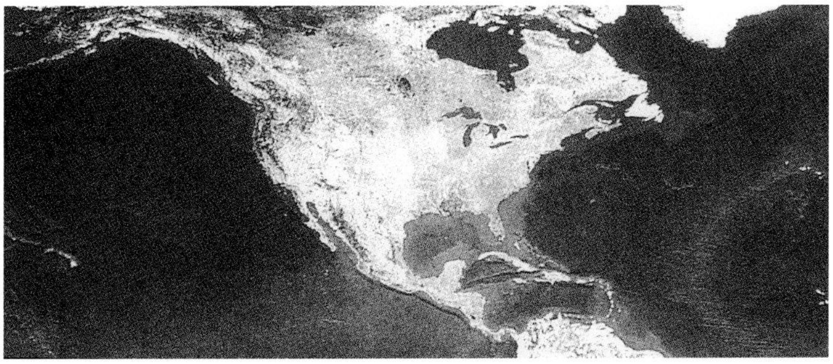

North America. *Eugen Sänger's* plan was to hit New York City via the Pacific Ocean, San Francisco releasing its bomb near Lake Superior of the Great Lakes, and heading home across the Atlantic Ocean as its bomb slowly glided down and self-detonated 2,000 to 3,000 feet above New York City.

ments and tilts of control surfaces. Other information also could be obtained, such as:

- The distribution of air pressure and temperature on the rocket bomber's surface could be determined through the use of thermo-couple elements for all operating velocities;
- Measurement of the location and magnitude of the forces acting on the aircraft;
- Direct tests of the reaction of stability by means of a *Cardan* suspension of the moving model at its center of gravity;
- Studies of the reaction of the rocket exhaust on flight conditions ... especially on the aero- dynamic forces and the stability of the wings and control surfaces;
- Takeoff and free flight characteristics of self-stabilized or remotely-controlled models, especially during the passage of the vehicle through the range of speeds from values below the speed of sound to values above;

Sänger believed these boosted vehicle tests might present special problems, not from the perspective of construction of the track itself, but rather in the frictional behavior of lubricated surfaces moving in contact with one another at speeds up to 5,000 kilome-

New York City in the early 1940s...target zero was Manhattan.

This first American atomic bomb was dropped on Hiroshima. It was 28 inches in diameter, 10 feet long, weighed 9,000 pounds, and with a destructive ability was equal to 20,000-tons (40 million pounds) of high explosives. The scientists were so sure that *"Little Boy"* would work that it was not tested before it was dropped over Hiroshima.

A *Boeing B-29* (*Eddie Allen*), of the type used to carry the world's first atomic bomb, seen moments before starting its takeoff run. This weapon had been nicknamed *"Little Boy."*

The atomic cloud over Hiroshima, morning of August 5th 1945.

Another *Boeing B-29* like the one pictured (*Eddie Allen - Boeing's* senior test pilot who died in the crash of a prototype *B-29* when it caught fire) carried America's 2nd atomic bomb over the Japanese city of Nagasaki on August 9th 1945.

Approximately 4.7 square miles in the center of Hiroshima was destroyed by the world's first atomic bomb on August 5th 1945.

America's 2nd atomic bomb was nicknamed "*Fat Man*." It was a plutonium type that was detonated over Nagasaki. It was five feet in diameter, nearly 11 feet long, and weighed 10,000 pounds. More advanced than the "*Little Boy*," and although its yield was the same (20,000-ton) its blast effect was greater.

ters/hour. Where tow tests, ordinary wind tunnel tests, or other methods of extending theoretical constructs were not possible, *Sänger* suggested the application of well-known plasma flow methods to the problems.

Phase 6 – Construction of a Prototype Airframe

Sänger wrote that once a preliminary decision regarding the shape of the rocket bomber had been made, based on wind tunnel and booster track tests, he could begin construction of the fuselage, wings, control surfaces, and important subassemblies, such as the pilot's cabin, fuel tank installations, and auxiliary equipment. When

The atomic cloud resulting from America's 2nd atomic bomb over the city of Nagasaki...an industrial center on the west coast of Kyushu on August 9th 1945.

Below: The range of a *Sänger* winged reusable orbital rocket bomber when being used in point bombing attacks with a rocket exhaust velocity of 9,842 feet/second [3,000 meters/second] and with bomb loads of 1- to 30-ton. Flights to the point bombing attack area would include two separate rocket burn times and an immediate turn around to home airbase after bomb release. The contour lines represent outer range of point bombing targets with bomb loads ranging from 1-ton, 5-ton, 10-ton, 15-ton, 20-ton, and 30-ton. For example, with a 1-ton bomb load, the *Sänger* rocket bomber could reach point targets up to 3,728 miles [6,000 kilometers] from its home airbase. At a 3,728 mile range this includes point targets in Central Africa, Hindustan, Eastern Siberia, the North Pole, Greenland, and a big portion of the North Atlantic. However, with a 30-ton bomb load, this *Sänger* rocket bomber's range is much shorter, but its destructive power is huge. A *Sänger* rocket bomber carrying 30-ton of bombs would have sufficient range to destroy cities in western France, western England, southern Sweden and Denmark, and western Poland, as well as hit Allied supply convoys on their way to the USSR via the North Sea.

Left: Nagasaki—after the smoke cleared, an area 2.3 miles long and 1.9 miles wide was totally leveled.

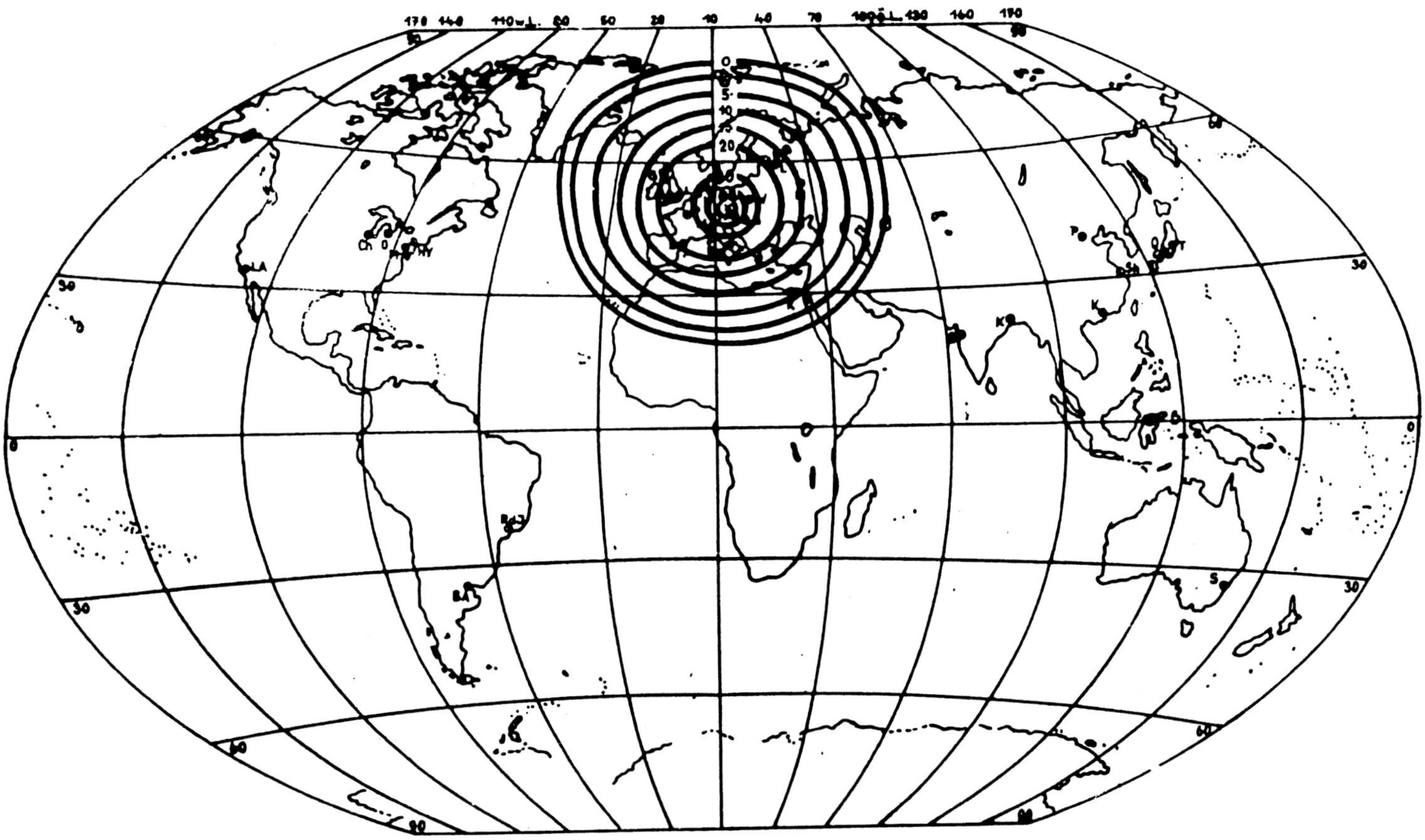

Cities of more than
● one million.
x Home base

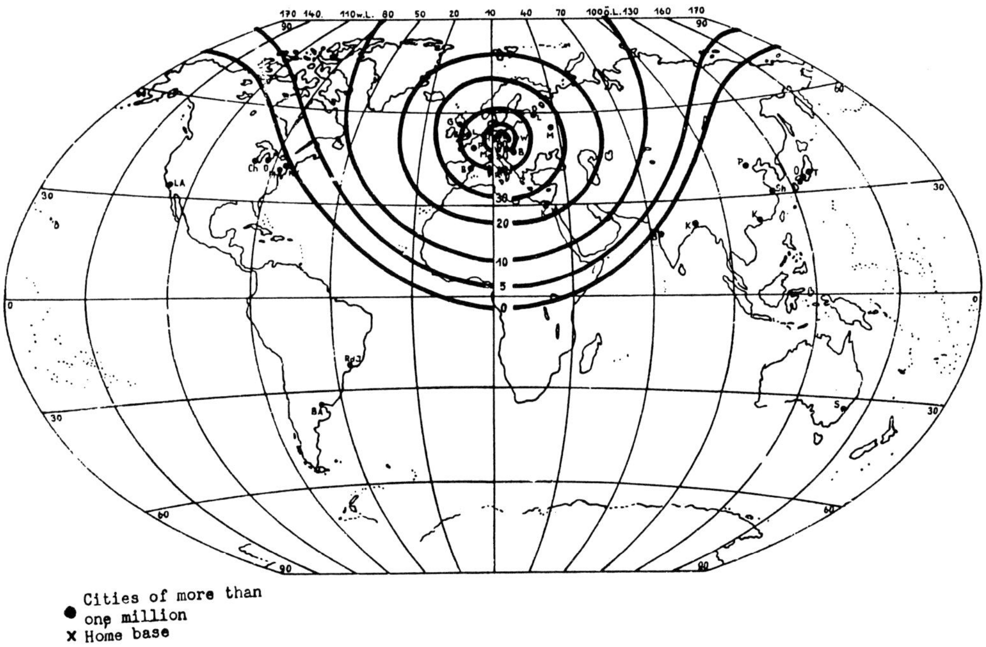

The range of a *Sänger* piloted, winged, reusable, sub-orbital rocket bomber when being used in point bombing attacks with a rocket exhaust velocity of 13,123 feet/second [4,000 meters/second] and with bomb loads of 5- to 30-ton. These flights include two separate rocket burn times and an immediate turn around to home airbase after bomb release. The contour lines represent the outer ranges of point bombing targets with bomb loads ranging from 5-ton, 10-ton, 20-ton, and 30-ton. Thanks to a rocket exhaust velocity of 13,123 feet/second [4,000 meters/second] the rocket bomber could deliver 5-ton of bombs for point bombing attacks near New York City, eastern Canada, most of the Atlantic Ocean, lower Africa, western India, and almost one half of the land area of the USSR.

considering the external shape, which was largely pre-determined by other requirements and constraints, *Sänger* said that air pressures – which would be far greater than those on ordinary aircraft – would be the decisive factor for construction of the fuselage, wings, and control surfaces. The dynamic pressure of the air on his rocket bomber, he calculated, could reach 3,000 kilograms/m^2 and would, therefore, lead to types of wing construction more commonly used in buildings or ships. *Sänger* wrote that challenges presented by requirements for a pressurized cockpit cabin already had been addressed in research on other advanced aircraft projects, so these data could be applied to his rocket bomber. He was thinking of the visionary work done by *Dipl.-Ing. Felix Kracht* at *DFS* in developing detachable pressurized cockpit cabins for the *DFS 228* and the *DFS 346* machines. *Sänger* admitted that there were no concepts on which to base tank installations for large quantities of fuel such as liquified gases, metallic suspensions, and so on. Nevertheless each compound had to be tanked in its own unique container for supersonic flight. *Sänger* mentioned that an extensive array of a whole new genre of flight instruments would need to be developed for his rocket bomber. In addition, auxiliary equipment would be required and included a variety of items such as:

The range of a *Sänger* winged reusable orbital rocket bomber when used in point bombing attacks with a rocket exhaust velocity of 16,404 feet/second [5,000 meters/second] and with bombs loads of 5- to 50-ton. These flights include two separate rocket burn times and an immediate turn around to home airbase after bomb release. The contour lines represent the outer ranges of point bombing targets with bomb loads ranging from 5-ton, 10-ton, 20-ton, 30-ton, 40-ton, and 50-ton. Thanks to a rocket exhaust velocity of 16,404 feet/second [5,000 meters/second] the rocket bomber could deliver 5-ton of bombs for point bombing attacks in all of North America, most of South America, areas north of Australia, all of Asia, and the entire USSR.

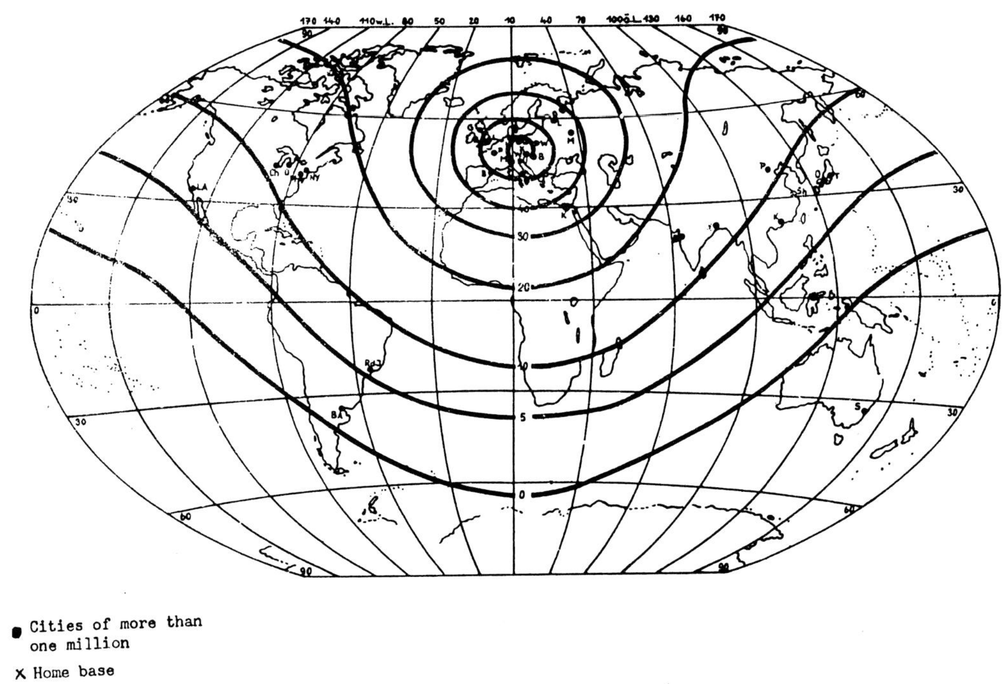

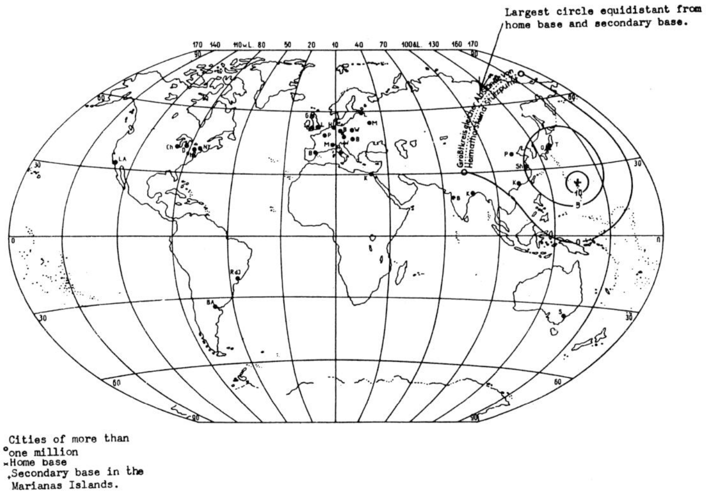

The range of a *Sänger* winged reusable orbital rocket bomber when used in a point bombing attack with a rocket exhaust velocity of 13,123 feet/second [4,000 meters/second] and the contour lines of a constant bomb-load and using an auxiliary landing airfield in the Marianas after the bomb attack. *Sänger* said that a point attack with two separate rocket burn times and the use of an auxiliary landing airfield was sensible only for those parts of the earth's surface which were nearer to the auxiliary landing airfield than to the rocket bomber's home airbase. The most far location would be a circle whose bomb attack points were equal-distance from its home airbase and the auxiliary landing airfield. The bomb lines would be limited to the region of the earth's surface beyond. The possible range of bomb attack in this case includes all of East Asia and a large part of the western Pacific Ocean. At a rocket exhaust velocity of 9,842 feet/second [3,000 meters/second] the rocket bomber's bomb load shrinks to small circles around the auxiliary landing airfield. However, for rocket exhaust velocities of 16,404 feet/second [5,000 meters/second] the possible range of bomb attack spreads out over practically the whole hemisphere opposite Europe so when using an antipodal auxiliary landing airfield, the winged rocket bomber's range would have been a staggering 12,428 miles [20,000 kilometers]. *Sänger* said that an example of a point bombing attack with two separate rocket burn times and an auxiliary landing airfield was a 2-ton bomb attack on the locks of the Panama Canal with an auxiliary landing airfield on the American west coast. Total distance covered by the rocket bomber would have been 7,860 miles [12,650 kilometers]. Total elapsed time between takeoff and landing would have been less than two hours [113.5 minutes].

- Acceleration and velocity meters that could measure speed outside the Earth's atmosphere;
- Altitude meters (non-barometric);
- Instruments for aircraft attitude control and steering;
- Navigation (near space);
- Precision aiming of bombs from very high altitudes;
- Optical instruments for aid in long-range approaches and high attack angle landings;
- Special controls for throttling or shut-down of the rocket engine after a partial fuel burn;

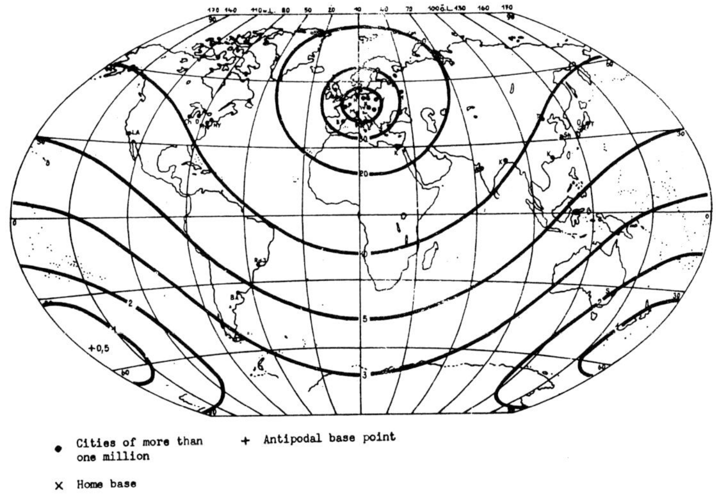

The range of a *Sänger* winged reusable orbital rocket bomber when used in a point bombing attack with a rocket exhaust velocity of 9,842 feet/second [3,000 meters/second], a single rocket burn time, and sacrifice of the bomber after the bombing attack. The winged rocket bomber would accelerate briefly after the bombing attack and its pilot would, at the end of the acceleration, parachute from the aircraft and allow it to destroy itself in its crash down. It was believed that the pilot would land a few miles [kilometers] away from the point of impact of his bombs and be captured. The contour lines represent bomb loads in tons ranging from 30- to 0.5-ton. In all cases, the attacking range covers the entire surface of the earth at 3,000 meters/second and a bomb load of 0.5-ton. At a rocket exhaust velocity of 13,123 feet/second [4,000 meters/second] the same rocket bomber could carry 8-ton of bombs. The bomb load would increase to 17-ton with a rocket exhaust velocity of 5,000 meters/second [16,404 feet/second]. *Sänger* believed that this procedure was naturally suited to unpiloted use of his winged rocket bomber.

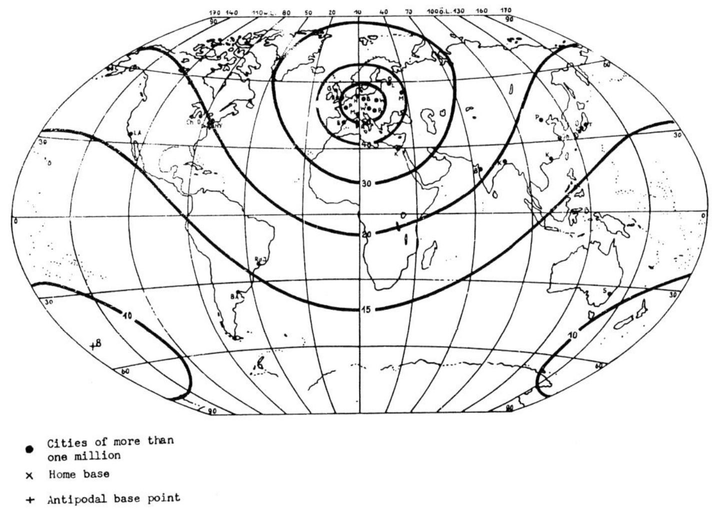

The range of a *Sänger* winged reusable orbital rocket bomber when used in a point bombing attack with a rocket exhaust velocity of 13,123 feet/second [4,000 meters/second], a single rocket burn time, and sacrifice of the bomber after the bombing attack. At 4,000 meters/second rocket exhaust velocity, the winged rocket bomber would be capable of delivering 8-ton of bombs. The contour lines represent bomb loads in tons ranging from 50- to 10-ton and the rocket bomber could reach any point bombing target anywhere on the surface of the earth.

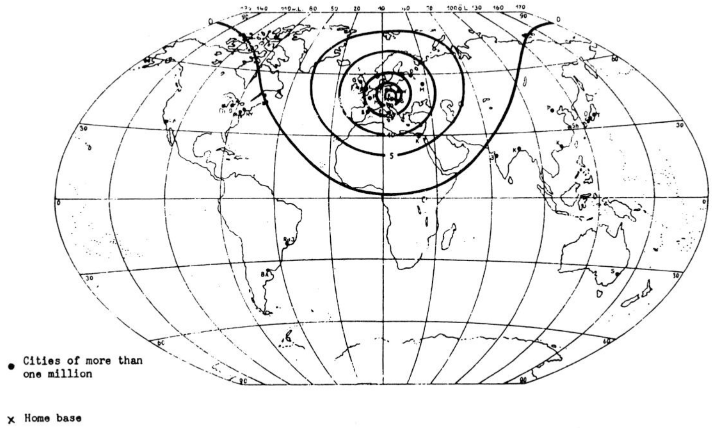

The range of a *Sänger* winged reusable orbital rocket bomber when used in an area bombing attack with a rocket exhaust velocity of 9,842 feet/second [3,000 meters/second] and a 10-ton bomb load. For comparison, the range of the winged rocket bomber with a rocket exhaust velocity of 13,123 feet/second [4,000 meters/second] and a 5-ton bomb load is given. The contour lines represent a constant 5-ton and a 10-ton bomb loads. With a 5-ton bomb load, full turning would be part of this area bomb attack procedure. The areas of attack are bounded by a closed curve which deviates from a circle because of the earth's rotation. Within range would have been the North Pole, Newfoundland, central Africa, and central Asia. The areas of attack widen considerably with a rocket exhaust velocity of 13,123 feet/second [4,000 meters/second]. At this exhaust velocity a 10-ton bomb load is possible so that the bombing ring expands so that now only Australia, the South Polar regions, the South Pacific, and the southern tip of South America do not lie within it.

Phase 7 – Static Bench Testing the 100-ton Rocket Motor Within the Airframe

Sänger wrote that bench testing of his 100-ton rocket motor represented the last stage before starting the actual flight tests. Static testing would check the satisfactory operation of the rocket motor under conditions where all the parts had been assembled into a rocket bomber's airframe. *Sänger* wanted to see the exact relationships between the thrust forces and the centers of mass and lift of the rocket bomber to determine and avoid any instabilities caused by the incorrect placement of his rocket engine. In addition, a side benefit of static testing within the machine's airframe would allow the pilot to get accustomed to some of the peculiarities of the new aircraft.

Phase 8 – Development and Testing Takeoff Monorail and Launching System

As far as the several kilometers long takeoff launch monorail, *Sänger* saw very few problems. He said the arrangement of its rails presented a relatively simple engineering and construction challenge and he foresaw no special problems. On the other hand, design of the takeoff sled was more formidable. Under a load of several hundred tons, it would have to accelerate in a very short time to 1.5 times the velocity of sound and then be slowed down even more rapidly. This, he said, would require special development and test-

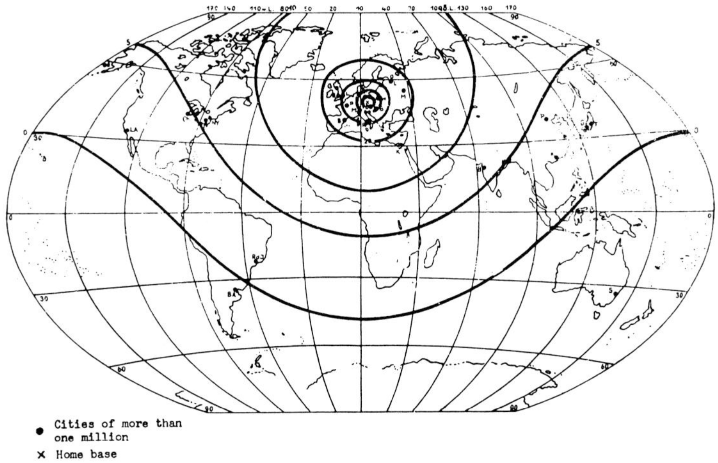

An example of an area bomb attack with a single rocket burn time and full turn would be a bomb attack on New York City at a range of 4,039 miles [6,500 kilometers] from central Germany. For this example, *Sänger* chose a 6-ton bomb and a rocket exhaust velocity of 13,123 feet/second [4,000 meters/second]. The rocket bomber would leave its catapult/launch monorail. Thirty six (36) seconds after launch and at about 75 miles [12,000 kilometers] distance from its takeoff point, the pilot would turn on the rocket motor and it would proceed to consume 84-ton of fuel within the next 336 seconds [5 1/2 minutes]. Nearing the end of its climb, the winged rocket bomber would have reached a velocity of 14,249 miles/hour [6,370 meters/second] at an altitude of 56.5 miles [91 kilometers]. At this point the *Sänger* winged rocket bomber has flown a distance of 457 miles [736 kilometers] from its point of takeoff and weighs 16-ton. Using only its store of potential and kinetic energy, the winged rocket bomber flies on to its target and releasing its bomb load 590 miles 950 kilometers] from the target. The winged rocket bomber would immediately go into an 180 degree turn spiral of 621 miles [1,000 kilometers] in diameter until it has reached the direction for its return flight to its home airbase. During turning, the winged rocket bomber's altitude has declined substantially in order to develop the aerodynamic forces required for the turn. At the end of its turn path, the rocket bomber's velocity is 8,277 miles/hour [3,700 meters/second] and its stationary altitude is 23.5 miles [38 kilometers]. The supersonic glide path back to home airbase is over 3,387 miles [5,450 kilometers] and time to home airbase from the bombing target is approximately 43 minutes and ends 62 miles [100 kilometers] from home airbase. The entire bombing flight has taken almost 80 minutes.

Right: A pen and ink drawing by *Sänger* featuring a distribution curve of bomb hits over a map of Manhattan in an area attack against a target point in the center of New York City. In order to destroy Manhattan to a 2nd degree destruction, the surface of the city would have had to received between 420- to 5,000-ton of bombs depending on their velocity of impact. *Sänger* calculated that the number of winged rocket bomber flights necessary for dropping this amount of destructive energy on Manhattan would be 84, providing that each rocket bomber carried 60-ton of bombs and their impact was low. If, on the other hand, the winged rocket bomber's velocity was 26,246 feet/second [8,000 meters/second] and each bomber released a single 1-ton bomb, then the number of winged rocket bombers needed would be 420 machines.

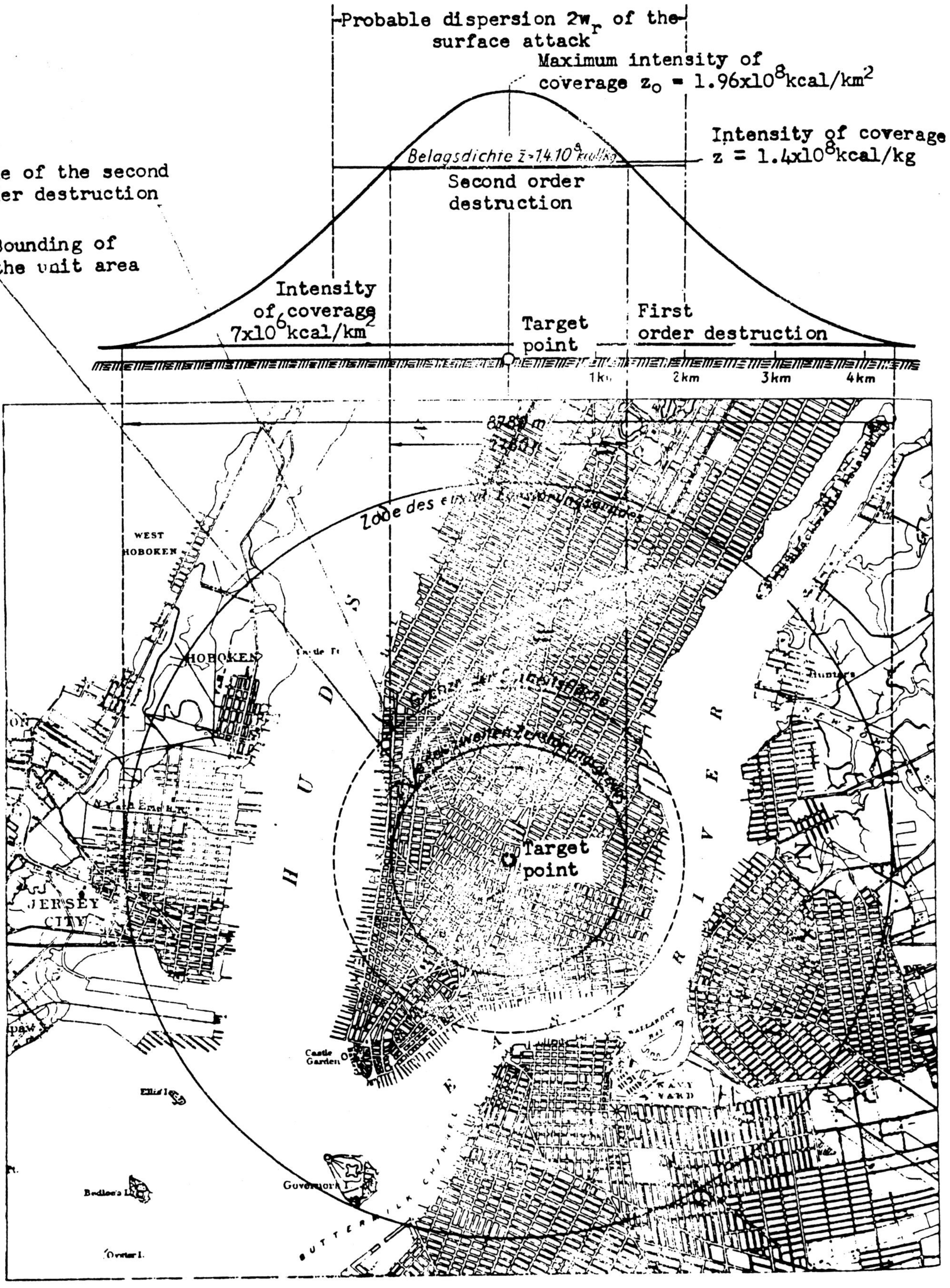
Probable dispersion 2w_r of the surface attack
Maximum intensity of coverage z_0 = 1.96x10^8 kcal/km^2
Intensity of coverage z = 1.4x10^8 kcal/kg
Belagsdichte z = 1.4.10^8 kcal/kg
Zone of the second order destruction
Second order destruction
Bounding of the unit area
Intensity of coverage 7x10^6 kcal/km^2
Target point
First order destruction
1km
2km
3km
4km
WEST HOBOKEN
HOBOKEN
JERSEY CITY
Castle Garden
Ellis I.
Bedloe's I.
Oyster I.
Governors I.
H U D S O N
E A S T R I V E R
BUTTERMILK CHANNEL
NAVY YARD
Target point

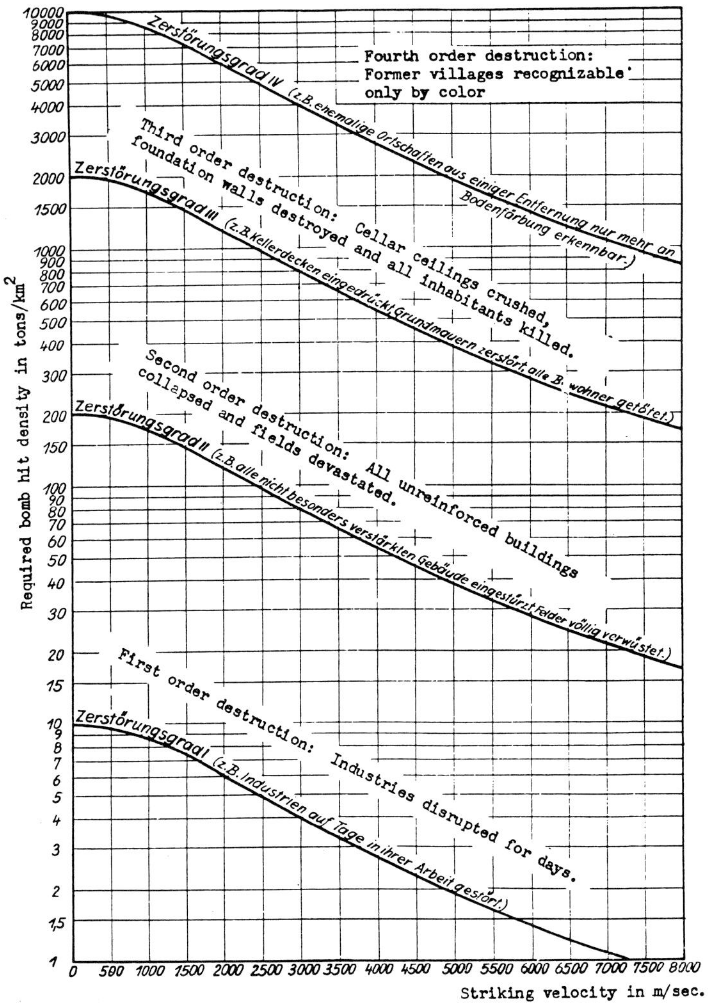

Left: *Sänger's* calculations for a required bomb hit density in tons on New York City to produce 1st through 4th order destruction as a function of the striking velocity of the bomb.

- **1st order destruction - industrial installations completely out of operation for several days;**
- **2nd order destruction - destroys cities so that all except specially reinforced buildings collapse and only cellars, basements, and foundations are usable;**
- **3rd order destruction - destroys cities so that even cellars are mashed in, all people inside the bomb area are killed and only foundation walls remain standing;**
- **4th order destruction - flatly destroys cities that their location is no longer discernable against the background;**

According to *Sänger*, as the striking velocity of the bomb in [meters/second] increased so did its destructive power.

ing. Other concerns about the sled launcher included its extraordinarily powerful array of booster rockets; friction of the sliding contacts of the sled riding on the rails of the launch track; and what means should be used for rapidly braking this launching vehicle. *Sänger* said a completely reliable operation of the takeoff monorail launch track would have to be carefully tested by catapult trials with dead weights comparable to the weight of the rocket bomber.

Phase 9 – Takeoff and Landing Tests

Sänger suggested takeoff and landing tests could be conducted with a very small quantity of fuel on board, with the bomber maneuvering on a taxi-strip with as long a length as possible. Testing would begin with a practically empty rocket bomber. After a very brief "burn" of its own rocket motor, reaching just a sufficient velocity to lift off, the bomber's rocket would be shut down and the vehicle would then come down again on its landing gear. This would not differ much from later landings after long flights. *Sänger* wanted

The *Sänger* winged reusable orbital rocket bomber shown over the east coast of the United States (New York City) with its bomb bay doors wide open exposing, for example, a single 5,000 pound radiation-spreading bomb. Digital image by *Dan Johnson*.

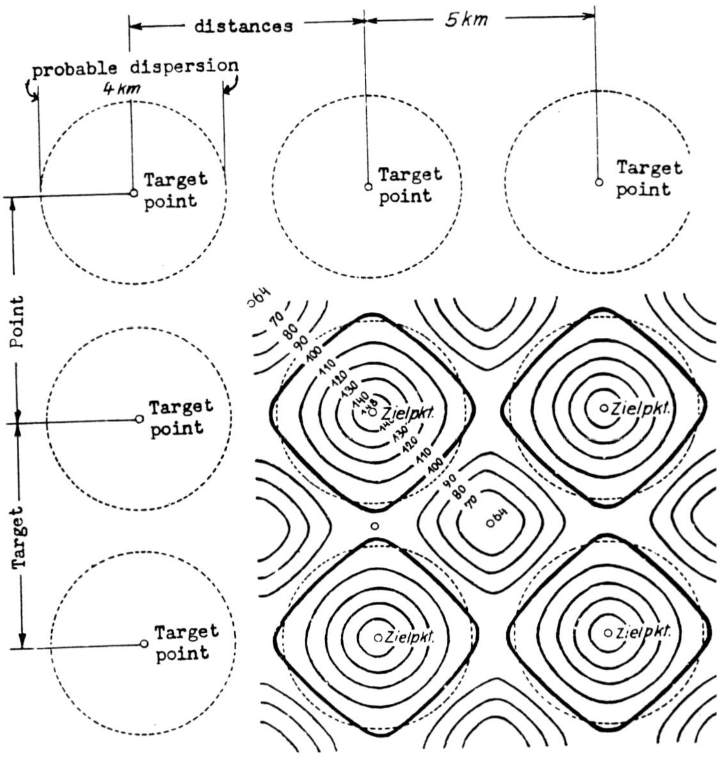

If larger connected surfaces than the unit surface are to be attacked then several points of the target can be aimed at so that the individual *Gauss* error curves partially overlap somewhat as featured in this *Sänger*- drawn illustration which shows the distance between aiming-points was chosen so that the average density over the whole surface is "*Z*" while the local densities are shown as contour lines. Thus, what we see is target distribution in the case of attack on very large surfaces and contour lines of equal bomb hitting density in percent of average value.

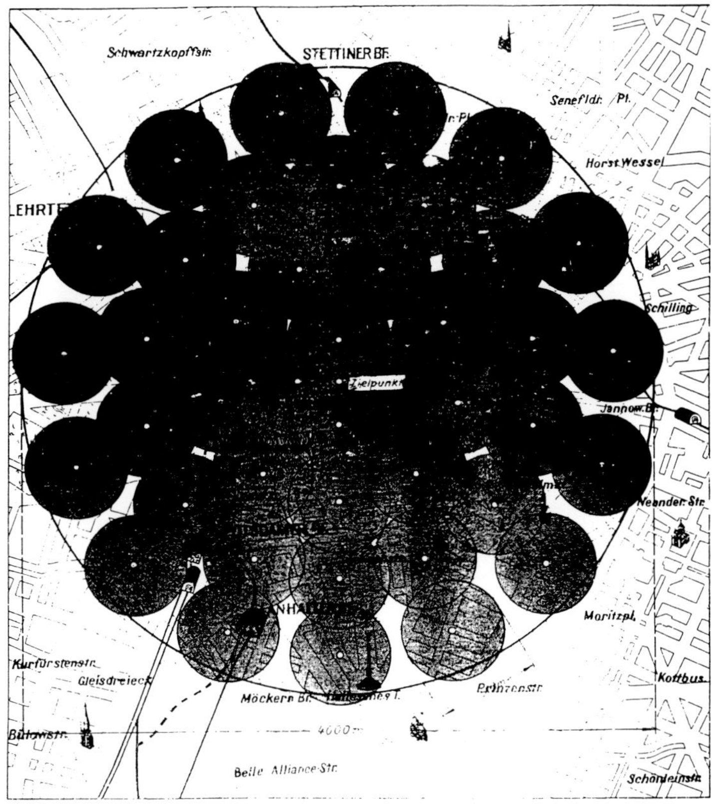

A poor quality image of the ideal *Gaussian* bomb hit pattern *Eugen Sänger* referred to in his wingned reusable orbital rocket bomber proposal. Shown is the idealized distribution of hits, according to the laws of probability overlaid on a map of the city of Berlin. *Sänger* assumed that 84 bombs of 60-ton weight for a total tonnage of 5,020-ton were dropped on the aiming point with low impact velocity. The resulting destruction would have amounted to 2nd order destruction.

these takeoff tests carried out in a large space so that landings could occur immediately after takeoffs so the machine would not have to turn around in the air. He wanted to avoid maneuvers in the air which might endanger the machine when its flight characteristics were still largely unknown. Only after many test flights and landings, when the bomber's flight and rocket power characteristics were better understood, would the test program engineers and the pilot attempt higher flights with more fuel in the tanks. The aircraft could be brought to somewhat higher altitudes, but still below 1,000 feet, then glided back down to the landing strip. Thus essential flight characteristics at low speed would be determined and improved such as:

- A better understanding of the best arrangement and actuation of the various control surfaces;
- The trimming, stability, and manoeuverability of the vehicle would be throughly checked;
- The airworthiness of the machine in all flight attitudes could be evaluated;

Sänger wrote if all his tests were satisfactory, then actual flight trials would be extended to higher altitudes and flight speeds. This phase of the trials would begin first by launch track tests on a very long launch rail so the machine wouldn't become airborne, but would only be accelerated and then be immediately slowed down again. Those "pre-launch" tests of the actual vehicle moving at gliding speeds of 800 to 1,800 kilometers per hour [497 to 1,119 miles per hour] were to be devoted especially to rechecking and affirming the results of similar launch track tests on smaller scale models concerning:

- Pressures;
- Temperatures;
- Aerodynamic forces;
- Stability;
- Vibrations;
- Effects of the driving rocket motor;
- Studies on the behavior of installed instruments;
- Studies on the behavior of installed apparatus;
- Studies on the behavior of the rocket motor under the influence of high accelerations and velocities;

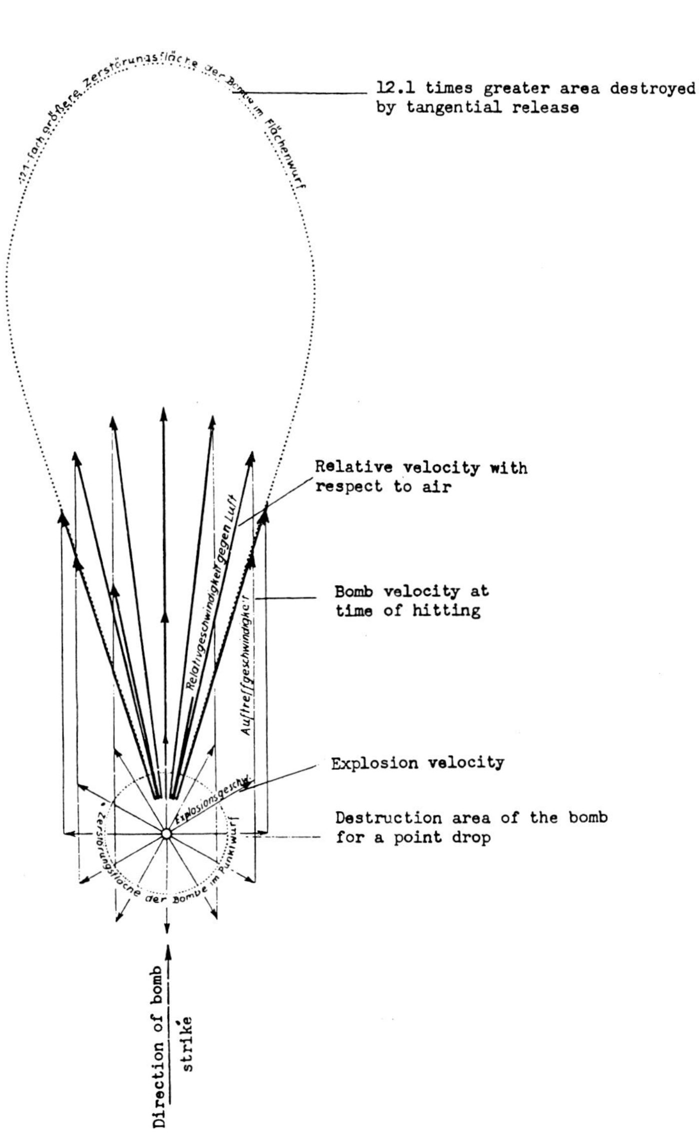

Sänger studied the mechanics of the explosive process. For example, he noted the area of destruction by an explosive charge was proportional to the energy available for the explosive wave. He found that the destruction area for similar bombs of equal tonnage but varying only in tangential velocity, a bomb thrown at the target by his proposed winged reusable orbital rocket bomber at the rate of 17,895 miles/hour [8,000 meters/second] would provide a 12.1 times greater area destroyed than did a direct point hit from a free-falling bomb. This is graphically shown in the pen and ink illustration. Shown is the destruction circle from a direct hit from a free-falling bomb. Notice that the explosive force scatters in all directions, however, with a bomb thrown at 17,895 miles/hour, the explosive force goes out in the direction of its flight. *Sänger's* piloted, winged, rocket bomber would be throwing its bombs, therefore, destructive areas would be larger than from a conventional bomber.

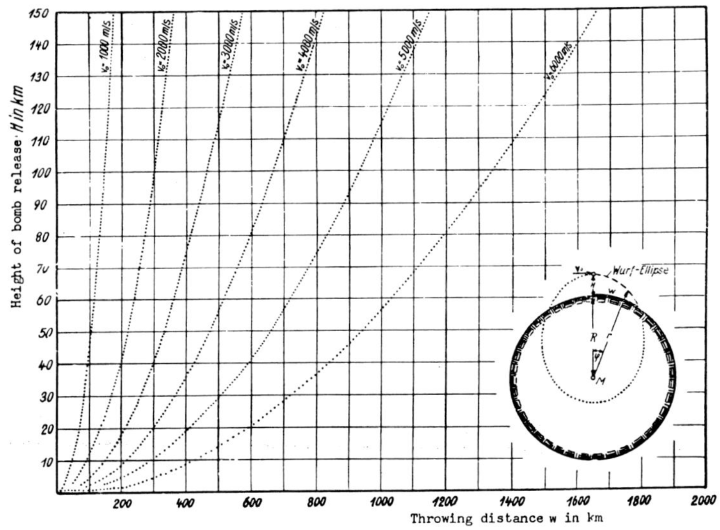

A pen and ink illustration by *Sänger* showing the throwing distance of bombs released by his winged reusable orbital rocket bomber at various speeds of meters/second. The range is 2,237 to 13,421 miles/hour [1,000 to 6,000 meters/second]. The higher the speed thrown at a target on or near the surface, the greater will be the bomb's destructive abilities and this fact was not lost on *Sänger* when he was seeking financial support to develop his winged rocket bomber.

- Development of suitable safety devices;
- Getting the pilot accustomed to the new phenomena of high accelerations, high velocities, special engines, and the new arrangements of the rocket bomber aircraft;

Phase 10 – Flight Testing the Bomber

Flight tests in which the rocket bomber would be catapult-launched, then actually be released from the launching sled, would follow next. Initially these catapult tests would begin with a limited quantity of fuel on board and the pilot allowing the rocket motor to op-

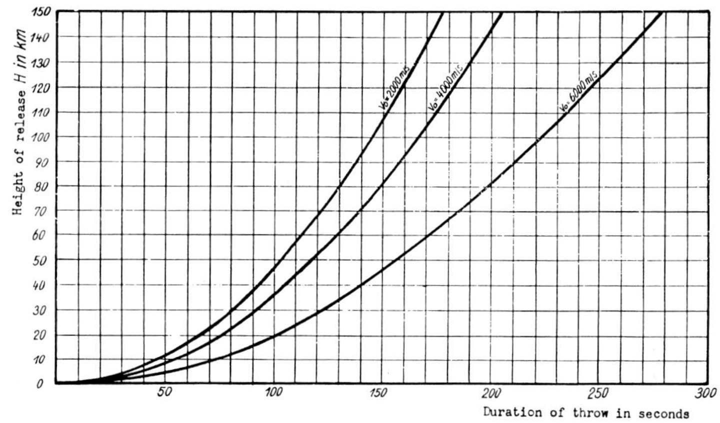

A pen and ink illustration by *Sänger* featuring the duration, in seconds, of a bomb throw by his wing reusable orbital rocket bomber from a release height of 93 miles [150 kilometers] and forward speeds of 4,474, 8,948, and 13,421 miles/hour [2,000, 4,000 and 6,000 meters/second]. At a height of 93 miles and a forward speed of 13,421 miles/hour, a bomb released by the winged rocket bomber would take approximately 4 minutes to hit its surface target.

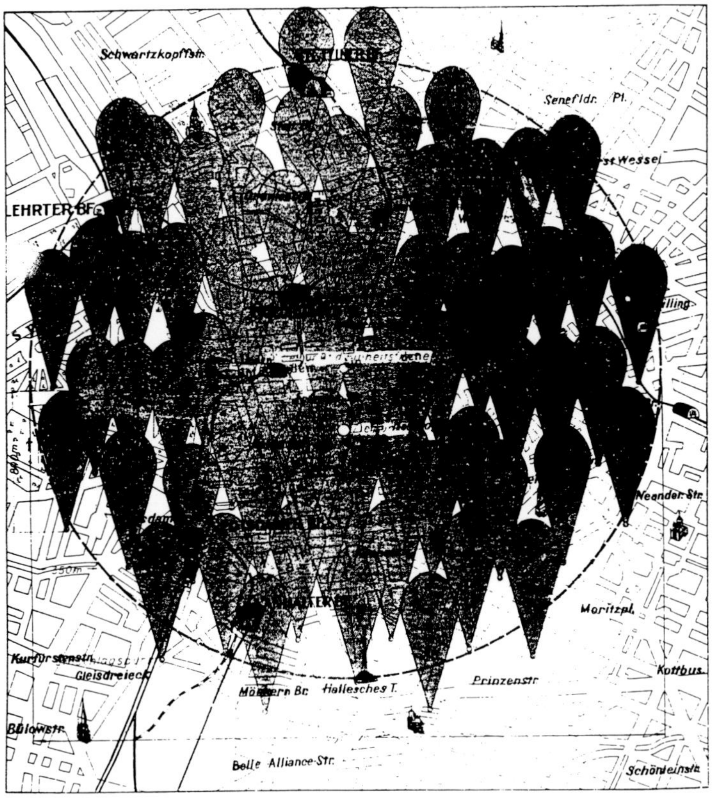

Sänger's example of an ideal *Gaussian* hit distribution of 140 separate bomb releases on an overlay map of Berlin at 8,000 meters/second [17,895 miles/hour]. Each release represents 3-tons of bombs for a total tonnage of 420-ton and the complete destruction of Berlin. Compare this small tonnage to the 5,020-ton of bombs dropped on an overlay map of Berlin at low velocity. Taking the speed advantage of *Sänger's* winged reusable orbital rocket bomber, it was clear that his winged rocket bomber could deliver a 2nd order destruction with fewer than 10% of the bombs required for a low velocity attack.

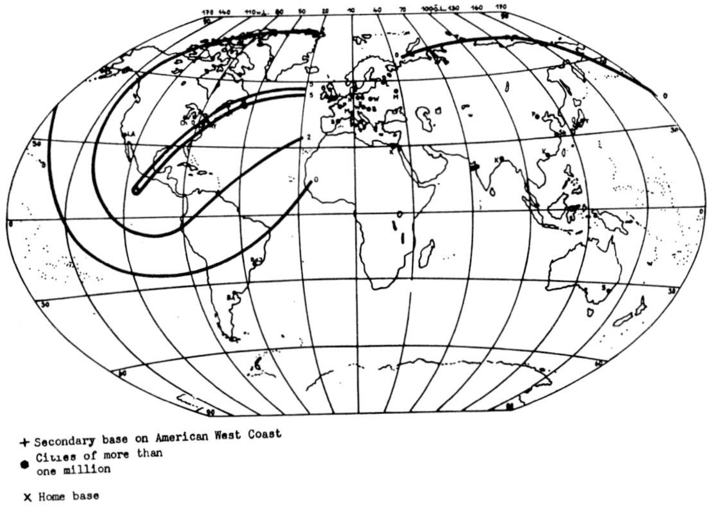

The range of a single *Sänger* winged reusable orbital rocket bomber when used in an area bombing attack with a rocket exhaust velocity of 9,842 feet/second [3,000 meters/second]. The contour lines represent a constant 5-ton bomb load and a constant 10-ton bomb load, partial turning, and the use of an auxiliary landing airfield on the west coast of America. The bombing procedure would begin with a single rocket burn time. In essence, the single rocket burn time would give the winged rocket bomber all the power which it would require until landing at a predetermined auxiliary landing airfield. The winged rocket bomber would release its bomb load in front of the target at high altitude and flight velocity and then carry out a turn at its existing high speed immediately after the bomb release. The rocket bomber would then head in the direction of the auxiliary landing airfield and glides there on its residual energy and lands. *Sänger* believed that the winged rocket bomber's pilot would be able to release his bomb load at any point enroute without altering the bomber's range provided that the bomb release would occur after its rocket motor was turned off. The range of this area bombing with a partial turn would be anywhere between takeoff and the auxiliary landing airfield. The pen and ink illustration features a 5-ton bomb load with 9,842 feet/second rocket exhaust velocity, however, with a smaller bomb load the possible area of attack stretches overall North America and considerable portions of South America and the Pacific Ocean.

erate for longer and longer periods on successive flights. Gradually higher velocities and longer climb paths would be conducted until the rocket bomber soon surpassed velocities attained only the fastest aircraft in the *Lufwaffe*. This type of "envelope expansion" flight testing would continue up to altitudes never previously reached. Takeoff speeds also would be increased with each successful test flight. *Sänger* said the most dangerous and difficult part of any flight test program is the realm where flight conditions move well beyond those known at the present and exhaustively validated by experience...to a new research path taken into completely uncharted aeronautical territory. *Sänger* said that this territory had to be conquered before moving on to *Phase 11*. Flight testing would involve the experimental test pilot taking into account the high altitudes; accelerations and temperatures; the effect of the forces on the wings and controls of the aircraft; testing of the aerodynamic heating of the aircraft's fuselage walls at various points of the aircraft; maintenance of stability and maneuverability beyond the speed of sound, and so on. *Sänger* suggested this stage of flight tests would have to be done very carefully, with a step-by-step increase in velocity. The reason? At very high altitudes, low air densities and high speeds, any slight accident could lead to catastrophe, since it would be practically impossible to leave this radically advanced performance aircraft by parachute.

Carefully phased flight testing, *Sänger* noted, would continue up to speeds not yet reached even by projectiles – speeds about 6 times the velocity of sound. Finally, the actual trial flights would be completed with final performance tests. *Sänger* said that he needed to determine simple data such as:

- Takeoff speeds and distances;
- Landing speeds and distances;
- Climbing performance;
- Consumption of fuel;

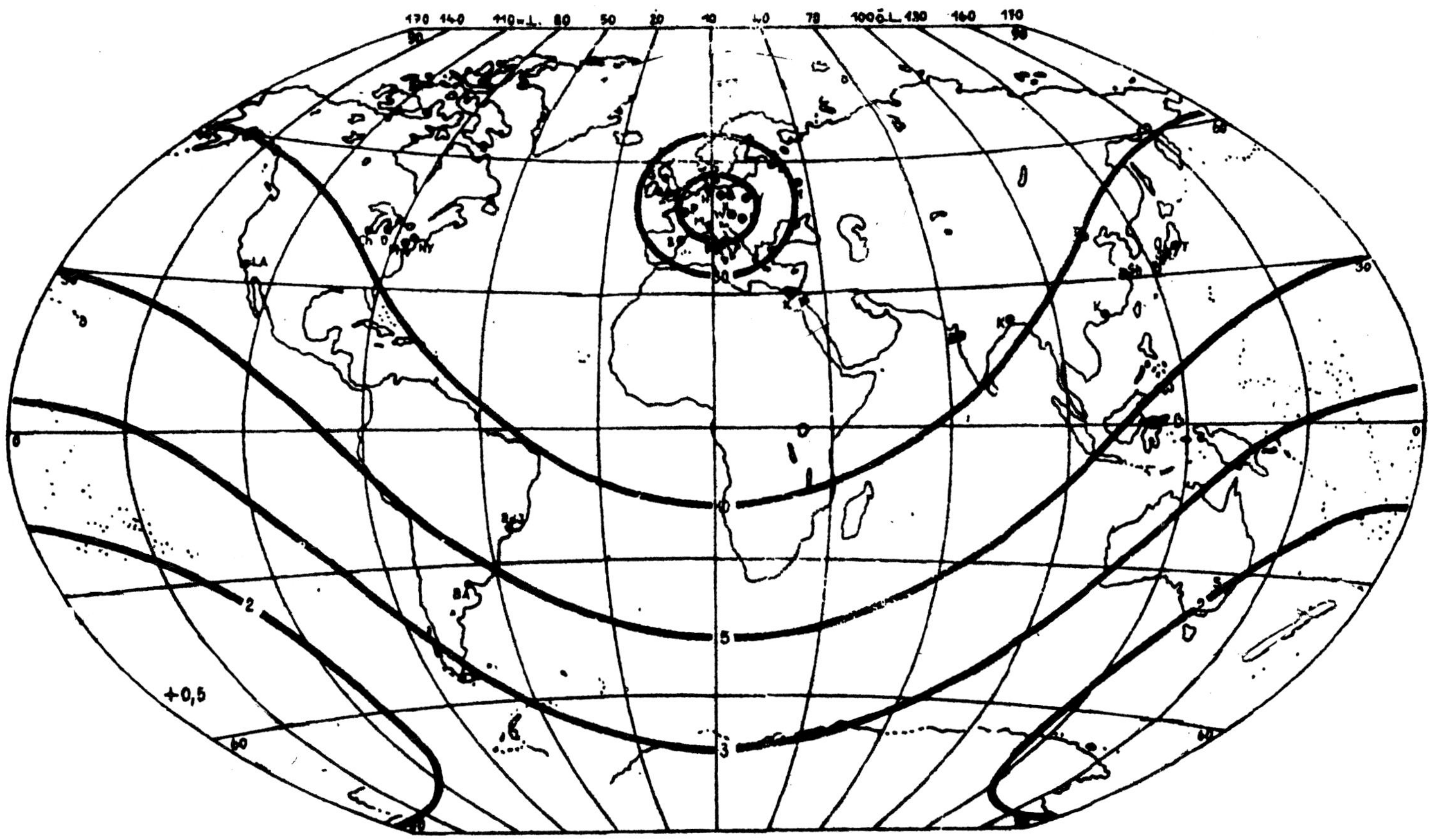

● Cities of more than one million

+ Antipodal base point

x Home base

The range of a single *Sänger* winged reusable orbital rocket bomber when used in an area bombing attack with a rocket exhaust velocity of 9,842 feet/second [3,000 meters/second]. The contour lines represent a constant bomb load of 3-ton with a single rocket burn time and circumnavigation of the earth. *Sänger* was attempting to show the technical superiority of his winged rocket bomber with a size and empty weight equal to that of a medium conventional bomber, could at a moderate rocket exhaust speeds, reach every point on the earth's surface with a bomb load of 3-ton and fly 24,900 miles [40,000 kilometers] all around the earth without an intermediate landing. An example of an area bombing attack with a single rocket burn time and circumnavigation *Sänger* used was a bombing attack on a city with a million population most distant from Germany...Sidney, Australia. In this case the range of attack would have been 10,253 miles [16,500 kilometers] and a possible bomb load of 3-ton. The flight would have gone as follows:

- **takeoff and motion after takeoff the would be the same as other winged rocket bomber launchings;**
- **thirty-six (36) seconds after takeoff the rocket motor is turned on and would consume 87-ton of fuel on board with in the next 348 seconds (5.8 minutes) as the winged rocket bomber(s) climbed to altitude;**
- **at the end of a 348 seconds (5.8 minutes) climb its forward velocity is 16,105 miles/hour [7,200 meters/second];**
- **the winged rocket bomber's altitude would be 62.8 miles [101 kilometers];**
- **the distance traveled from takeoff point would be 506 miles [815 kilometers];**
- **weight of the winged rocket bomber would be 13-ton;**
- **forward speed would drop to 671 miles/hours [300 meters/second];**
- **the winged rocket bomber would begin its supersonic long descent of 24,350 miles [39,185 kilometers] length;**
- **after traveling 6,214 miles [10,000 kilometers], the strongly oscillating descent would be damped sufficiently so that at the release point, 9,570 miles [15,400 kilometers] from its takeoff point, would be smooth enough at the stationary attitude to enable accurate aiming for its bomb release;**
- **at the bomb release point the altitude would be 30.4 miles [49 kilometers];**
- **the winged rocket bomber's forward velocity would be 14,316 miles/hour [6,400 meters/second];**
- **the range of bomb projection would be 684 miles [1,100 kilometers];**
- **after the bomb release the winged rocket bomber would began its supersonic glide back to its takeoff airbase and weighing only 10-tons;**
- **the rocket bomber would land at its takeoff airbase 13,060 seconds [2 hours 17 minutes] after leaving it earlier;**
- **the *Sänger* winged reusable orbital rocket bomber will have traveled 24,826 miles [40,000 kilometers];**

The *Sänger* winged reusable orbital rocket bomber at about 78/80 miles altitude. It is seen in its ricocheting flight, like a flat stone hitting the surface of a lake. *Sänger* claimed that the ricochet would throw the machine into layers unable to fully support it, however, some distance away the orbital bomber would hit the denser layers again and bounce off once more. The path, seen as a whole, would be a wavy line...something like a roller coaster...the undulations slowly diminishing in size, both vertically and horizontally.

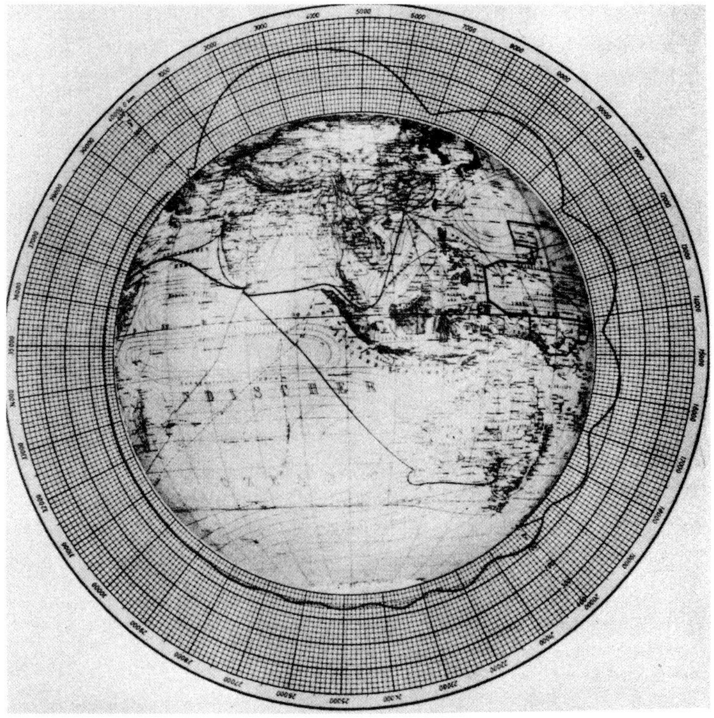

This is a pen and ink illustration *Eugen Sänger* used to describe his winged rocket bomber's skip-flight trajectory circumnavigating the globe to hit Sidney, Australia. It's dark line inside the grid above the globe, shows the peaks and troughs the rocket bomber with its 3-ton bomb load would experience after powering up to altitude, continuing on with its mission, and landing several hours prior back at its takeoff airbase in Germany.

Once the above flight characteristics had been completed, *Sänger* would then continue to determine the following performance characteristics:

- Maximum speed;
- Maximum operational altitude;
- Maximum range;

During this exhaustive testing of the rocket bomber's performance characteristics, *Sänger* would be particularly interested seeing how his machine reacted to surface temperature distribution on all leading edges as forward velocity increased. He had some sort of remote-reading thermometer system that could detect dangerous levels of aerodynamic heating resulting from air friction and stagnation. High frictional temperatures on leading edges of the wings and nose of the aircraft, might result in the slightest melting or deformation and subsequent blunting of these surfaces, which could then lead to steadily spreading molten regions, and the immediate burning up and disintegration of the whole rocket bomber. If performance flights were carried out until the expected limiting values of 7,000 meters per second velocity, 150 kilometers altitude, and 40,000 kilometer range were safely and successfully reached, then flight studies could be considered completed.

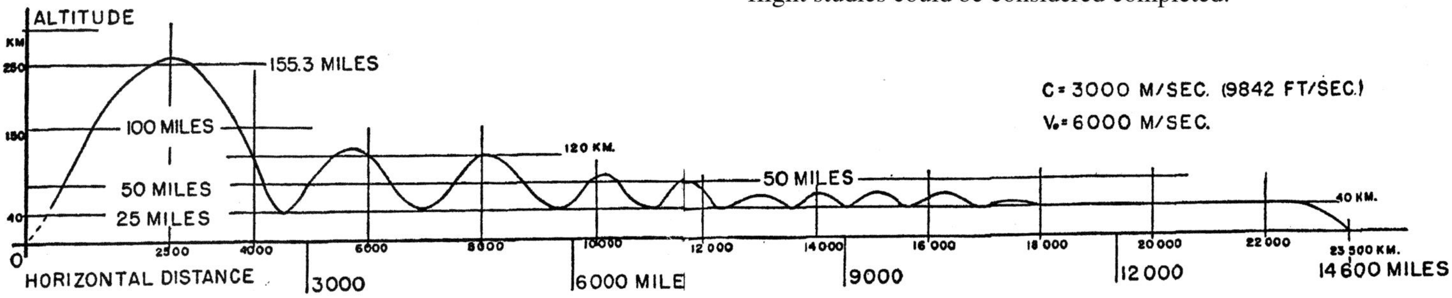

A diagram featuring a typical *Sänger* winged rocket bomber's flight path between antipodal points on earth...New York City and Berlin. New York City is not antipodal to Berlin. The opposite side of Berlin would be Australia/New Zealand. For example, the winged rocket bomber would climb to 155 miles altitude and over a distance of about 3,000 miles fall to within 25 miles of the earth's surface. At this point, the *Sänger* would skip off the dense atmosphere up to about 78 miles altitude for a total distance now covered of about 3,600 miles. It would have been at this point when the *Sänger* winged orbital rocket bomber would have bombed New York City. The *Sänger* would continue home heading across the Atlantic to Germany. Total powered flight time: 5 minutes. Total flight duration: about 2 1/2 hours. Total distance covered: approximately 14,600 miles [23,500 kilometers].

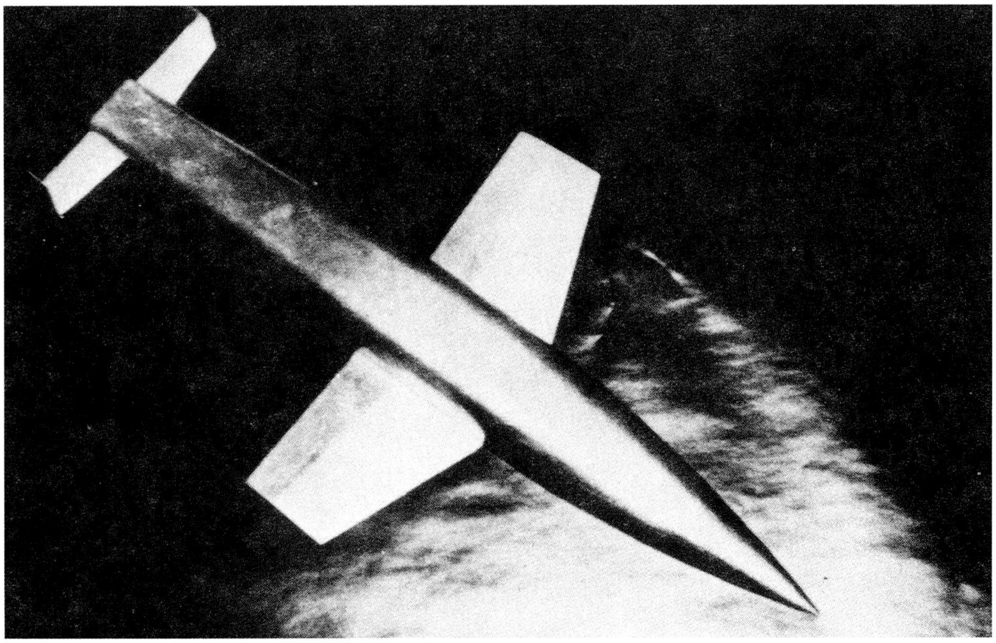

A wind tunnel scale model of the *Sänger* winged reusable orbital rocket bomber seen as if appearing in space with the earth in the background. The piloted, winged, rocket bomber is seen from its starboard nose side in the mouth of a wind tunnel at *DVL*.

***Phase 11* – Navigation Tests with the *Sänger* Rocket Bomber**

Sänger's next important task of development concerned the equipment and procedures required for successful low Earth orbital navigation after propulsion had ceased. These challenges were daunting — to be able to give the pilot an exact knowledge of the bomber's flight path, enable corrections of its course, and permit an exact calculation of the moment of bomb release required considerable navigational skill and speed. This precision navigation, *Sänger* believed, would have to be checked in a very large number of flight tests, since success of any bombing attack from space depended on the accurate and rapid performance of in-flight calculations and constant tracking of position of the vehicle in relation to the Earth's surface.

***Phase 12* – Bomb Release Trails**

This was the last planned phase of the research and development work on the rocket bomber. *Sänger* needed have practical verification of his preliminary theoretical work on the processes likely to occur during the fall of the bombs and their contact with the Earth. So far as the relation between point of impact and point of release from aircraft at various heights and flight speeds was concerned, *Sänger* felt the oceans provided a suitably large testing area. He believed a study of bomb impact on land would be more troublesome, since very large uninhabited areas would be required. *Sänger* suggested a few test areas in the Arctic regions or deserts, however. The realities of the war were becoming evident. As *Sänger* was writing his rocket bomber proposal in 1944, Germany's remaining available possessions throughout the world having sufficient open land areas for advanced weapons testing were being overrun by Allied armies.

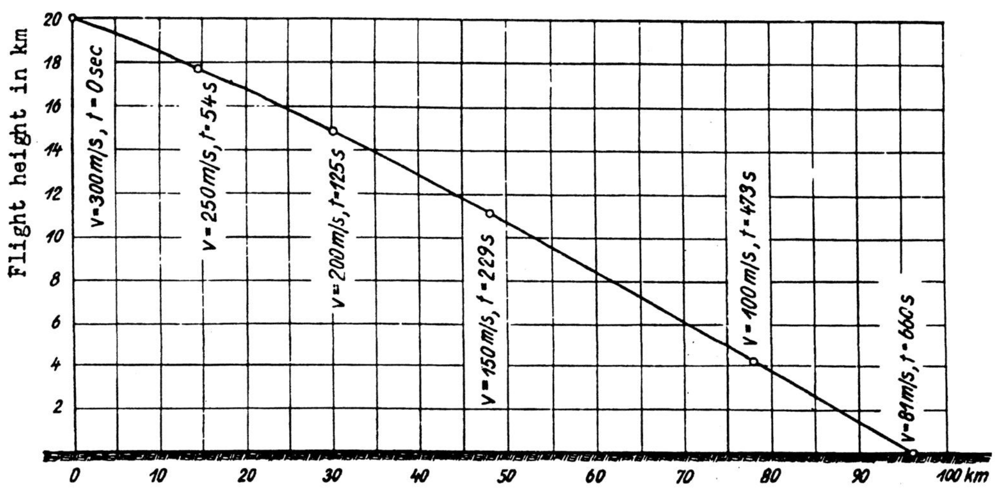

A pen and ink graph by *Sänger* featuring the typical subsonic descent path which would have been taken by his empty winged reusable orbital rocket bomber returning to either it home airbase or an auxiliary landing airfield. The 10-ton winged rocket bomber's subsonic gliding flight would began at a forward velocity of 984 feet/second or 671 miles/hour [300 meters/second]. At this rate the subsonic descent to the surface of the earth would last 11 minutes and would end near the surface of the earth at a velocity of 179 miles/hour [288 kilometers/hour] when landing can occur.

Reaction to Sänger's Orbital Rocket Bomber Proposal

It appears that *Sänger's* enthusiasm embodied in the draft did not receive the same degree of interest from officials at the *RLM* as it did from Western countries and the USSR. On March 17th 1942, the conclusions of *Sänger's* draft were rejected outright by the *Luftfahrtforschungsanstalt-Hermann Göring.* A bitter argument followed. It has been suggested that the closing-down of *LFA*-Trauen/Fassberg in the Summer of 1942 was due to fuel shortages, as well as objective and personal differences between *Sänger* and his immediate superior, the director of *LFA*. Personnel differences and fuel shortages did in fact occur, but the main reason for the closing of *LFA* was that research and testing at Peenemünde had produced such discouraging results that *Göring* saw no necessity to continue operation of a costly, duplicative and competitive development program. At the time of *LFA's* closure, *Sänger* and his research team were working on a liquid oxygen/diesel fuel rocket engine giving four times as much thrust as that of Peenemünde Experimental Rocket Center's *V2* rocket missile's engine (50,000 pounds), and concepts to utilize liquefied ozone and metallic fuels had progressed to the laboratory stage. *Sänger* was ordered to stop all work on the development of his 100-ton thrust rocket motor. He and his research assistants would be allowed to continue their research work, but only on *Lorin* propulsive duct motors and nitric acid/diesel oil rocket engines, but they would have to do it elsewhere. The entire team was transferred to a Special Engine Department of *DFS*...a demotion, really, and one *Sänger* didn't deserve.

In August, 1942, *Sänger* and his top research assistant, *Irene Bredt*, left *DVL* for positions with the *Deutsches Forschunsanstalt für Segelflug*-Ainring (*DFS*) – the German Research Institute for Gliding Flight, located at Ainring. *Sänger's* other able research assistant, *Helmuth von Zborowski*, had left in early1941, accepting a position with *BMW*-Berlin/Spandau, thanks to *Reichsführer SS Heinrich Himmler's* influence. *BMW* had taken an interest in the liquid-fuel rocket for propulsive purposes as an offshoot of their research into the possibilities of boosting the performance of aircraft powered by piston engines. *Count Von Zborowski* also had been a reserve officer in *Himmler's* secret police organization. By mid 1942, *Himmler* had penetrated and established a powerful presence within Germany's rocketry research and development environment. *Himmler* had been closely following *von Zborowski's* career while he was a research assistant with *Sänger* at Trauen/Fassberg. *Eugen Sänger's* boss at *DVL* (*Busemann*) suggested that *Sänger* apply to become a member of the *SS*...that it would be a

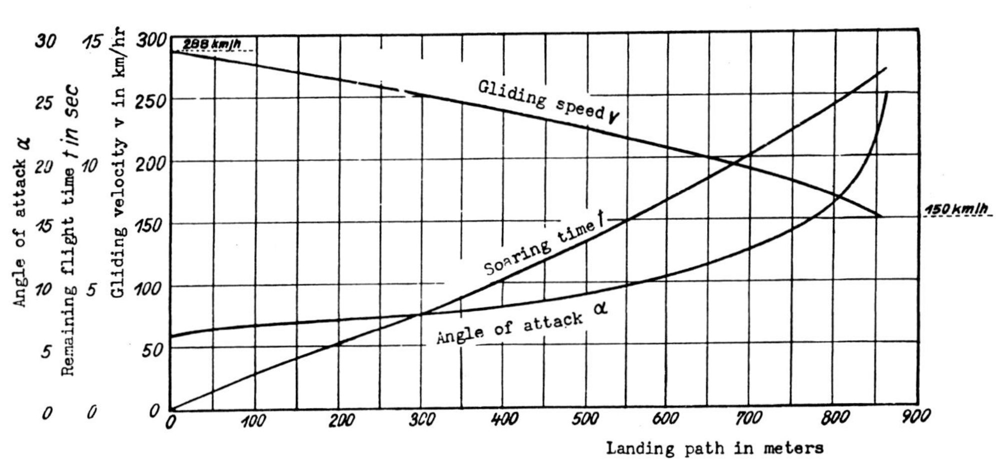

A pen and ink graph by *Sänger* featuring the angle of attack for all velocities of his winged rocket bomber between 179 and 93 miles/hour [288 and 150 kilometers/hour] for the 10-ton empty bomber. *Sänger* wanted his winged rocket bomber to touch down at 93 miles/hour [150 kilometers/hour] which he claimed was required by military glide-landings.

Eugen Sänger **believed that his 10-ton empty weight winged rocket bomber, due to the machine's superior aerodynamics, high-speed landing, and lack of air resistance that a number of landing aids would be required to slow it down. Although the drag (braking) parachute had been in use by the *Luftwaffe* when *Sänger* proposed his design such as several *DFS* gliders, *Dornier Do 217s*, and the twin turbojet powered *Horten Ho 9 V2* all-wing and so on, it appears that he didn't consider the use of one for his rocket bomber. Instead, *Sänger* would have turned his machine into a tail dragger (two individual tail wheels) so that its fuselage and wing's leading edge would bite into the air, slowing the machine as it rolled to a full stop.**

The first *FIST* ribbon braking parachute developed in *Nazi* Germany. It is shown installed on a *Dornier Do 217 M-1* and mounted in its tail cone.

Testing the *FIST* ribbon brake parachute on a *Dornier Do 217 M-1*. The apparatus appears to be fully deployed.

The *Sänger* winged reusable orbital rocket bomber is landing. Its main wheels have touched down on a wet, hard surfaced landing strip leaving tire smoke and water vapor clouds swirling behind the machine. Although the winged rocket bomber appears to have a tricycle landing gear it really didn't...it was a tail dragger. The nose wheel was included to keep the machine from nose-diving, but its use was not intended. Digital image by *Josha Hildwine*.

The starboard side rocket exhaust thrust end of the *Sänger* winged reusable orbital rocket bomber as seen from ground level and featuring its two widely separated tail wheels. The outer rocket exhaust nozzles would have been used for maneuvering at ultra high altitudes. Digital image by *Mario Merino*.

good idea. *Sänger* had joined the Austrian *Nazi Partie* between 1930-1931, and sought to become a member of *Himmler's SS* about the same time. Membership was denied on the grounds that he was politically unreliable. So it appears that *Sänger* played no role in the *Nazi SS* between 1930 and 1945, held no rank nor much he benefitted from any *SS* connection. But one thing is obvious; *Sänger* associated with known *SS*-connected scientists in the field of rocket research such as *General Walter Dornburger, Wernher von Braun*, and others. *SS* officer *Helmuth von Zborowski*, on the other hand, continued his pioneering research work into nitric acid/diesel fuel rocket engines at *BMW*. These two liquids became hypergolic (self-igniting) when mixed in a rocket's combustion chamber and the result was a powerful, 'instant' thrust. This technique was used to add a 'boost thrust' capability to the *BMW 003R* gas turbine engine by installing a small rocket engine adjacent to the turbine's exhaust cone.

Zborowski had investigated some 6,000 potential fuel combinations which had demonstrated the utility of using nitric acid as a chemical bearer of the oxygen required for fuel consumption. By 1943, *Zborowski's* research resulted in creation of the *BMW 718*, a small liquid bi-fuel rocket that produced 2,755 pounds thrust for 3 to 5 minutes, combined with the standard *BMW 003A-1* turbojet engine of 1,764 pounds thrust. This power plant combination became the *BMW 003R*. Field testing of this unit was underway at war's end on specially modified *Messerschmitt Me 262A1* fighters and *Arado Ar 234C* four turbojet powered bombers. Earlier, in 1941,

***Sänger* wrote that his winged rocket bomber was to be a tail dragger so that as the machine settled down on its tail wheels the aircraft's angle of attack would catch the wind and act as a brake, thereby slowing it down. In order for the winged rocket bomber to have a high angle of attack it would have had to have very long oleo struts. Shown is a *Sänger* winged reusable orbital rocket bomber with a ridiculously high angle of attack thanks to some extra long oleo struts. This was an unusual approach because at the time *Eugen Sänger* was designing his winged rocket bomber, drag parachutes were in use to slow aircraft down after landing. The *Arado Ar 234B* and the *Horten Ho 9* are two such examples. Digital image by *Mario Merino*.**

The *Sänger* winged reusable orbital rocket bomber as seen from its starboard rear and featuring its extra long main wheel oleo struts. Digital image by *Mario Merino*.

Helmut von Zborowski, his research chemist *Hemesath*, *Wernher von Braun* (1912-1977) and his rocket engine specialist *Walter Thiel*, were investigating the use of a nitric acid/diesel fuel rocket engine for *von Braun's* proposed *A-8* rocket. The *A-8* would be an improved version of the *A-4* more commonly known as the *V2*, which had a normal range of 156 miles from its 68,500 pound thrust liquid oxygen/alcohol engine. *Von Braun* wanted to power his *A-8* rocket with a proposed 33-ton (66,000 pounds) thrust hypergolic engine to give the vehicle a range up to 280 miles [450 kilometers. The *A-8* concept was canceled in mid-1942, due to political and engineering reasons. Nevertheless, had the war lasted longer, the *BMW* rocket engines of *von Zborowski* certainly would have been formidable competitors to the *HWK* rocket engines of *Helmuth Walter* for aircraft propulsion applications.

A starboard side nose on view of *Eugen Sänger's* winged orbital rocket bomber. It does not appear that *Sänger* intended to use a drag parachute upon touch down to slow the machine. Instead, *Sänger* wrote that his winged rocket bomber would be a tail dragger featuring two widely spaced tail wheels. *Sänger* reasoned that the machine's high angle of attack would bite into the air slowing the machine. Digital image by *Mario Merino*.

Eugen Sänger and DFS-Ainring

DFS-Ainring was headed by *Professor Walter Georgii*. It appears that *Georgii* arranged for *Sänger's* draft report on a rocket-engined intercontinental space transport, now re-titled a rocket engine for intercontinental bombers, to be published. The size of the original draft was cut in half for publishing and it came to be known as *UM 3538*, a "secret command report" of German Aviation Research. *Sänger* received official approval to release his report on September 22nd 1944. Approximately 125 copies of the highly secret report were published; about 70 copies were actually distributed. The Red Army found only 3 copies of *Sänger's UM 3538* – one copy was uncovered while sacking *RLM* offices in Berlin; another copy was recovered at *Siebel Flugzeugbauwerk* in Halle; and one copy was found at Peenemünde. American Intelligence found other copies in the files of *DFS*-Ainring. *Eugen Sänger* was then 39 years old.

Germany's unconditional surrender came on May 8th 1945. *Eugen Sänger* and his long time assistant *Irene Bredt* fled Germany together, emigrating to France and taking employment as consulting engineers at the *Arsenal de l'Aeronautique* at Chatillon. *Professor Walter Georgii, Sänger's* boss at the *DFS* also fled to France. *Count Helmuth von Zborowski* rejoined the *Sänger/Georgii* team in France and experimented with engines suitable for vertical take off machines. *Zborowski* constructed an experimental vertical take off machine known as the *SNECMA C-450 "Coleopter."* It was flight tested only once, but was involved in a serious crash, severely injuring its test pilot. The machine was not repaired. Although the climate and living conditions were pleasing enough, the *Sänger* team experienced many disappointments in their professional lives. Both *Eugen Sänger* and *Irene Bredt* published extensively, but the

The *Sänger* winged reusable orbital rocket bomber as seen from its starboard side. Notice that its nose wheel has been retracted up into its nose wheel well and the gear cover doors closed. Digital image by *Mario Merino*.

Oberst Siegfried Knemeyer, the acknowledged boss of the powerful "_club,_" composed of movers and shakers within the _RLM_ 1944-1945. _Knemeyer_ was desperately seeking a flying machine capable of bombing the United States in order to bring about a cease fire or at least a conditional surrender. Among the flying machines being considered were _Werhner von Braun's_ proposed piloted _A-9_ liquid rocket with its _A-10_ booster rocket, the _Horten_ brothers proposed all-wing _Ho 18A_ and/or _Ho 18B_, and _Eugen Sänger's_ proposed orbital rocket bomber. _Knemeyer_ is shown exiting the cockpit of a _Horten Ho 3D,_ a motorized sailplane. He was frequently piloting another _Horten_ all-wing ...the twin piston motored _Horten Ho 7_.

Two of _Siegfried Knemeyer's_ favorite people..._Walter Horten_ (left) and _Reimar Horten_ (right).In the background is a _Horten Ho 2_ (left) and a _Horten Ho 3_ (right). _Walter_ kept their motorized _Horten Ho 3D_ sailplane and the twin piston motored _Horten Ho 7_ at the Oranienburg Air Base just for _Knemeyer_ to fly any time he wished. Oranienburg was located in Berlin's northwest suburbs.

widespread reconstruction of the French aviation industry after the war provided no support or funding for far-reaching and expensive projects such as the former *Luftwaffe* rocket-powered orbital bomber. Besides, the frequent changes of Government in post-war France prior to the accession to power of *General Charles de Gaulle* severely limited continuity of current projects. Projects approved in the afternoon by the outgoing regime, often would be canceled the next morning by the new incoming government.

Sänger and *Bredt* were married in October, 1951, in France. They collaborated with the French on development of the turbojet/ramjet powered research airplane called the *Griffon*. The following year *Irene Bredt-Sänger* give birth to a son, *Hartmuth*. Then on April 19th 1952, *Eugen Sänger's* late-war friend *General Walter Dornberger*, the former Commander of the Experimental Station For Rocket Weapons at Peenemünde, and now associated with *Bell Aircraft Corporation* in America, asked *Sänger* if he and his family

The proposed six _Jumo 004B_ turbojet powered _Horten Ho 18A_ "_Amerkia Bomber_" designed expressly for _Siegfried Knemeyer_ in late 1944. Its wingspan would have been 131 to 138 feet [40 to 41 meters] and a loaded weight of 4-tons. Maximum speed anticipated was 559 miles/hour [900 kilometers/hour] and a range of 7,500 miles.

The huge bomb-proof hangar unfinished at war's end for _Nazi_ aircraft and turbojet engine manufacturing located near Muhldorf/Dachau. This is where the _Horten Ho 18A_ would be constructed according to _Albert Speer_, head of the Ministry of Armament. _Fritz Sauckel_, also from _Speer's_ Ministry of Armament, had assured the _Horten_ brothers that there would be sufficient labor from the numerous concentration camps nearby such as Dachau.

The size of this concrete structure was immense. The arrow seen near the bottom center of the photo points to the height of a man.

The heaviest bomb used by the *Luftwaffe* was their 5,511 pound [2,500 kilogram] *SB 2500 "Max."*

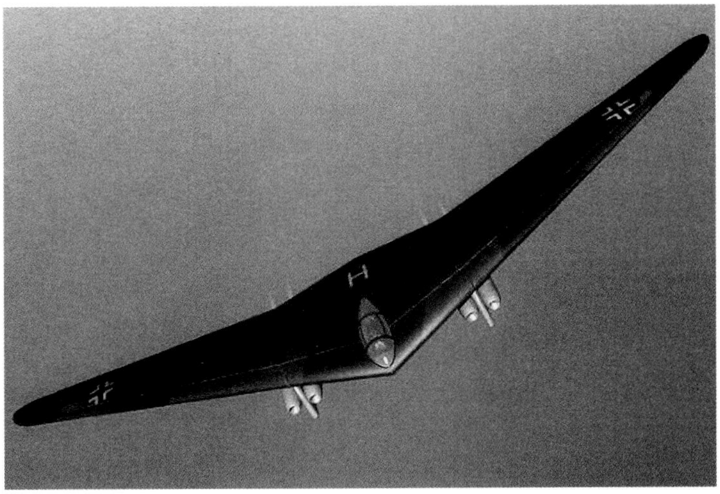

Reimar Horten had been assured of the availability of 2,866 pound thrust turbojet engines from *Heinkel- Hirth*...their *HeS 011A*s. *Reimar* reasoned that his all-wing *Amerika bomber* could get by with only four of the *Heinkel-Hirth HeS 011A* turbojet engines. He then redesigned the *Horten Ho 18A* into a smaller more compact flying machine known as the *Horten Ho 18B*. Its four *HeS 011A*s would be placed each side of a fixed yet skirted landing gear assembly. Its range was anticipated at 6,835 miles [11,000 kilometers] with a maximum level speed of 528 miles/hour [850 kilometers/hour] with a service ceiling of 52,492 feet [16 kilometers] attitude.

Right: This is how the *Nazi's* radioactive silica wrapped 5,000 pound bomb would have looked. It is shown zeroing in on Manhattan Island - New York City after being dropped by one of *Eugen Sänger's* winged reusable orbital rocket *Amerika Bombers*.

***Wehrmacht General Walter Dornberger* and *Wernher von Braun* shown at ceremonies when each were awarded the *Ritterkrauz* in recognition of their work on the *A-4* (*V2*) rocket. *Wernher von Braun* received the civilian equivalent of the award. These two men were responsible for their bi- liquid fueled *A-4* (*V2*) rocket missile. Although it had a range of less than 200 miles, *von Braun* and *Dornberger* were anxious to enlarge the basic *A-4* (*V2*) into an intercontinental bomber with a range of 3,000 miles. This range would be sufficient to deliver a bomb to New York City from a site in Western Europe.**

An *A-4* (*V-2*) test flight at a launch pad at the Peenemünde Experimental Rocket Center, which will carry it nearly 200 miles out to the Baltic Sea.

would like to emigrate to America and work with *Bell Aircraft* on an experimental *"X- plane"* later known as the *X-20* or *Dyna Soar. Bell's X-1* had been the first machine in the world to officially exceed the speed of sound in level flight. *Eugen Sänger* and his wife *Irene* chose not to accept *Dornberger's* offer; they remained in France where they had been living and working since war's end.

This was a fateful decision. Perhaps the *Sängers* could have avoided the embarrassment nine years later (1961) when *Eugen Sänger* choose to work as a rocket consultant for Egyptian *President Gamal Abdel Nasser.* It was alleged at the time that *Sänger* had assisted Egypt in the development and manufacture of combat rockets with the ability to strike Israel. Egypt did in fact have combat rockets, including the *Al Zafer and Al Kaher* and the Soviet *Sa-2* (*SAM*) ground-to-air missiles.

***Wernher von Braun's A-4* (*V2*) rocket didn't have the range to attack America. Although the *A-4* carried a warhead of 2,150 pounds its range was only about 210 miles.**

Sänger certainly was not the only German scientist consulting for Egypt. *Willy Messerschmitt* also was there. He had been consulting with the Egyptians and designing their supersonic fighter, the *Me Ha 300* – a plane designed to be equal or superior to the French-built *Mirage IV*, Israel's hottest aircraft fighter at the time. *Wolfgang Pilz,* formerly of the Peenemünde Experimental Rocket Center and an expert on rocket fuels, also had gone to France, and then to Egypt after the war. While in France, he and others developed France's *Veronique* and *Super-Veronique* rockets. Other Germans involved included *Paul-Jens Georcke*, an expert on guidance and controls and *Hans Kleinwachter*, an expert on guidance and electronics.

Sänger Appointed to the West German Research Institute for the Physics of Jet Propulsion

Sänger had been appointed in August, 1954, to direct and guide the *Forschungsinstitut für Physik der Strahlantriebe* (*FPS*) or "Research Institute for the Physics of Jet Propulsion," Stuttgart, West Germany. It was the first time since the end of *World War II*, that rocket research was being resumed in Germany after ten years of inactivity. In 1955, West Germany was granted the right to conduct experimental aeronautical and space research. The new *FPS* was to be concerned solely with the theory of rocketry. Its research staff would not be permitted to indulge in practical research. Typical research fields for the *FPS* included:

- theory of jets and rockets;
- physics of combustion;
- gas radiation;
- physics of wall effects.

The scope of *FPS* research was determined by the fact that modern development of aeronautics towards supersonic flight and astronautics primarily depended upon the research of more efficient jet turbine and rocket-powered propulsion. More than 10 years before this, Germany had a pre-eminent role in the development of high-speed, turbine-powered aircraft, *FPS* noted. Meanwhile, other countries had made significant technical advancements in some fields, especially in regard to turbojets, turbojets with afterburners, and liquid fuel rockets. Other types of jets such as ramjets, solid fuel rockets, take off booster rockets, and nuclear rockets were still in the early stages of research and development. *FPS*, by hiring the *Sängers*, hoped that work within the Institute would certainly contribute towards the solution of the physical problems of ramjet propulsion in the supersonic field, and also for the development of liquid rockets, nuclear liquid rockets, and photonic jet propulsion. *Sänger,* in his new position, also had clearance to see coincidental data on America's space program as well as intelligence information collected on the USSR's space efforts.

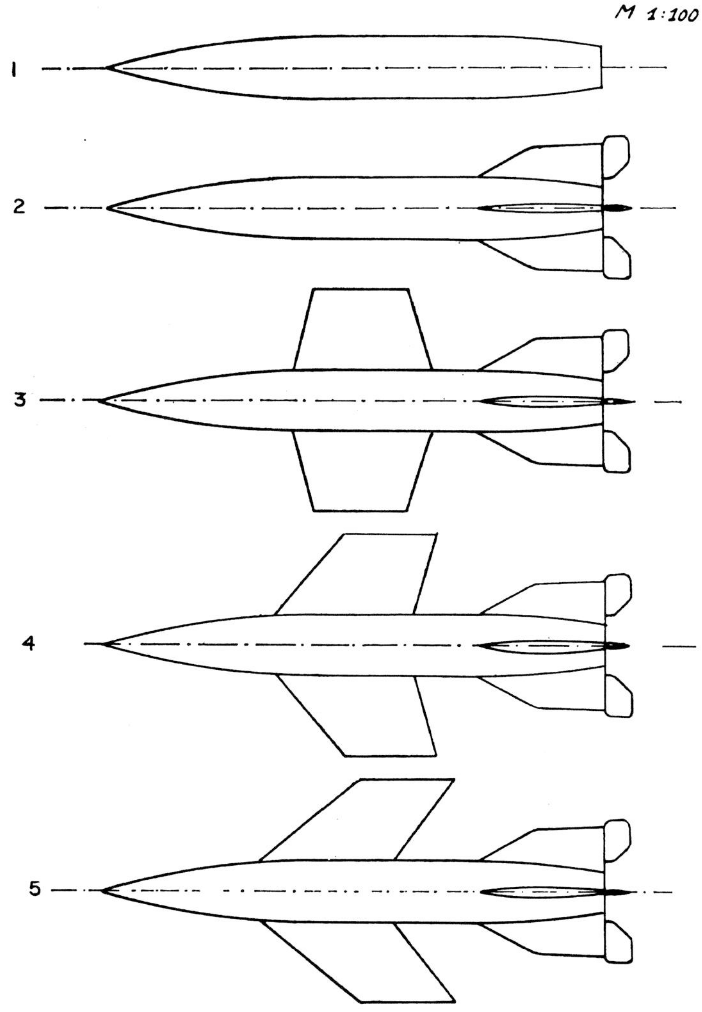

The evolutionary process leading up to the winged, piloted *A-9*. The most satisfactory performance was given by model #5 with wings swept back at an angle of 45 degrees to the fuselage.

The various aerodynamic shapes including wings...straight, wedge, and sweptback tested to determine the most efficient for the *A-5*. Configuration #4 was selected, and two prototypes were constructed prior to war's end.

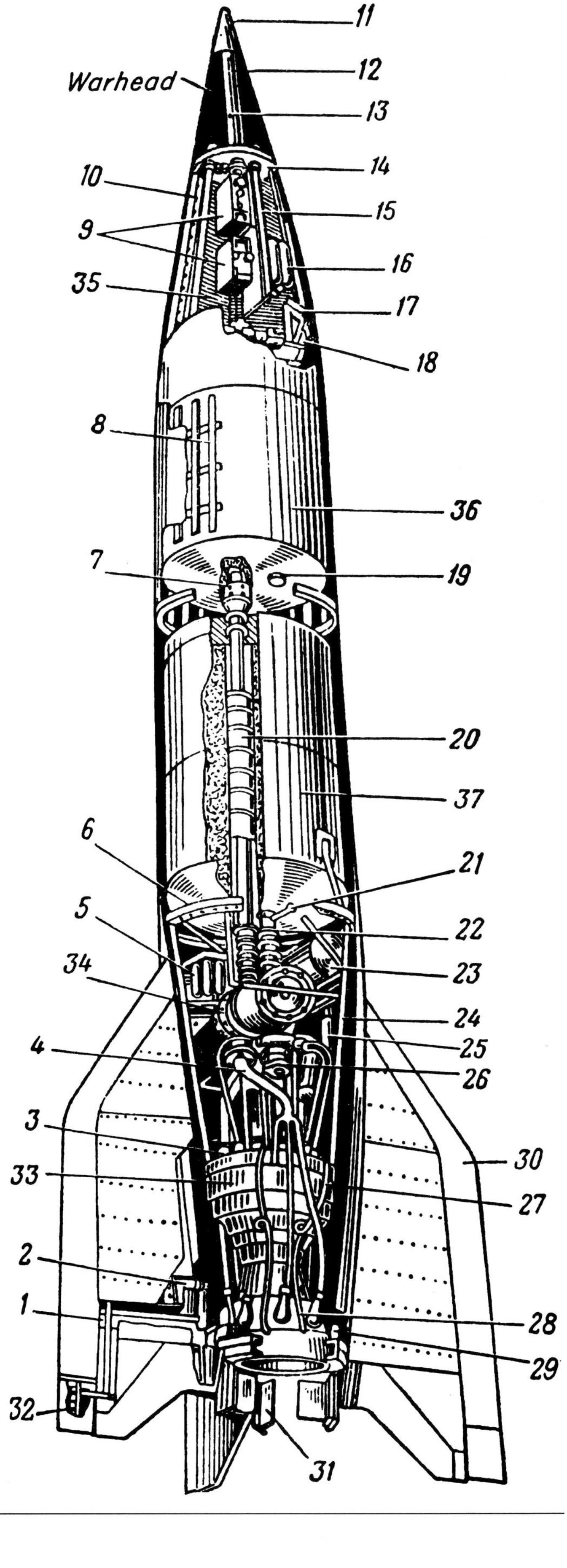

One of only one, or possibly two test flights made by the *A-5* prior to war's end.

Right: A pen and ink drawing of the standard *Wernher von Braun A-4* (*V2*) and featuring its component parts.

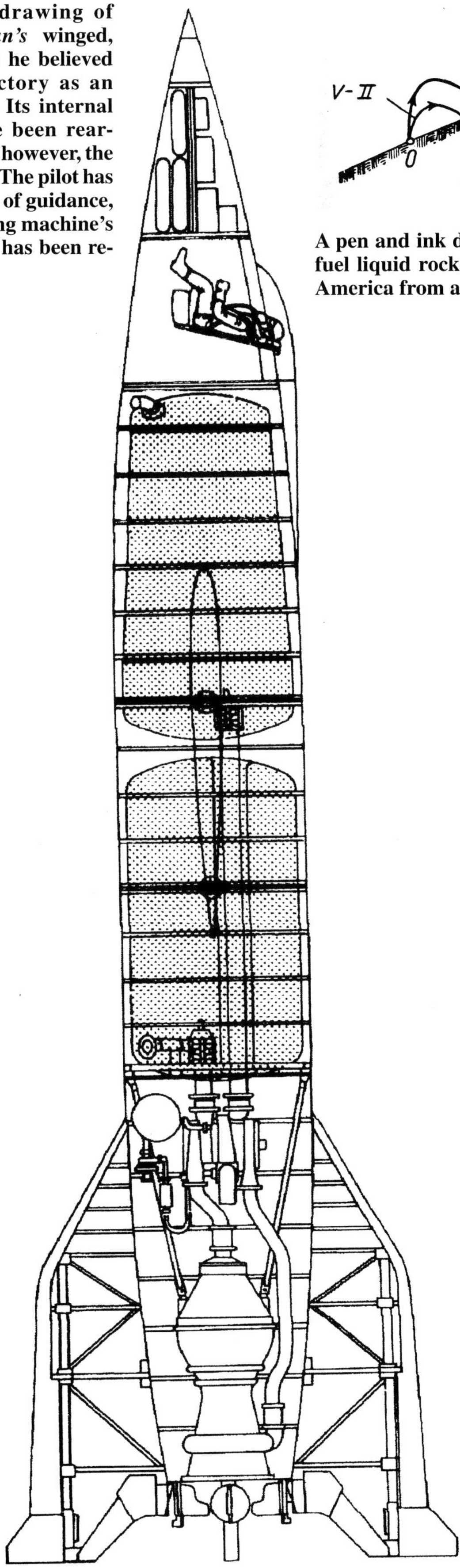

A pen and ink drawing of *Wernher von Braun's* winged, piloted *A-9* which he believed could be satisfactory as an *Amerika Bomber*. Its internal components have been rearranged somewhat, however, the warhead remains. The pilot has taken over the job of guidance, so most of the flying machine's internal guidance has been removed.

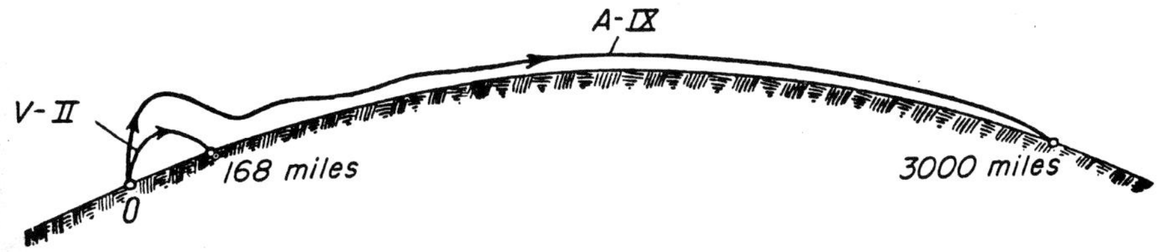

A pen and ink drawing featuring the flight path of the proposed winged, piloted *A-9* bi-fuel liquid rocket which is calculated to give up to 3,000 miles range...enough to reach America from a launch site in western Europe.

Sänger Fired from *FPS* for Consulting with Egypt in the Development of Rocket Missiles

Eugen Sänger was lecturing in the United States in November, 1961, when he was informed by newspaper reporters that he had been dismissed from his post by the Bonn Office of the West German Office for the Protection of the Constitution because he had exceeded his authority in serving Egypt. *Sänger* claimed that he had been assured that Egyptian rockets would be used only for atmospheric research. Later, after *Eugen Sänger's* death, *Irene Bredt-Sänger* stated that she and her husband left the Stuttgart-based Institute for Physics due to a disagreement over salaries.

Newspaper accounts that scientists from the Federal Republic of West Germany, several of whom were on the public payroll – like *Eugen Sänger*, were activity assisting the United Arab Republic with the manufacture of advanced offensive weapons outraged citizens and governments world-wide. Privately, the West German government had been aware of this collaboration for a long time. When news was publically disclosed that none other than *Dr. Eugen Sänger* was helping *General Gamal Abdel Nasser* perfect rockets for use in the destruction of Israel and other countries – as part of *Nasser's* aggressive designs on his neighbors – the world community reacted angrily.

***Wernher von Braun's* proposed piloted *A-9* bi-fuel liquid rocket would really have been a two-stage rocket. The first stage would have been known as the *A-10* booster rocket and it would have dropped off at about 2,000 miles down range.**

A pen and ink drawing of the proposed winged, piloted *A-9* along with its proposed booster *A-10*.

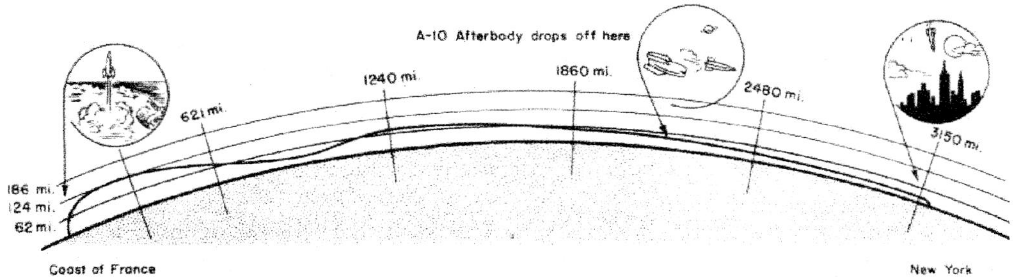

A pen and ink illustration of *Wernher von Braun's* proposed flight path from a launch site in Western Europe on its way to bomb New York City...some 3,150 to 3,500 miles away depending on its closeness/location to the Atlantic Ocean. Notice that the *A-10* rocket booster burns out, released, and left behind at about 2,000 miles down range.

A-9

A-10

The alleged collaboration between *Sänger* and his team of West German rocket scientists and *President Nasser* resulted in development of a powerful arsenal of military rockets whose planned use was the destruction of Israel, asserted U.S. *Congressman Halpern* from New York State. It appeared that Congress, as well as many American citizens, believed that West German rocket scientists like *Eugen Sänger,* all bearing West German passports, were in Egypt perfecting these rockets, and were doing so with the full knowledge and tacit support of the Bonn Government. Furthermore, it was alleged by *Congressman Halpern* that *Sänger* and others were perfecting military rockets in Egypt for the completion of *Adolf Hitler's* goal for the mass murder of the Jewish people. Egypt's closely guarded missile center, *Project 333* – where many of the West German's worked – was located near Cairo. An estimated 400 German scientists and technicians, most of them alleged to be from the wartime German *V-2* missile center at Peenemünde, were thought to have been working on the first Egyptian-made missiles with warheads possibly containing radioactive materials. *Sänger* was reportedly working on the following missiles:

- *al Zafar* - 375 kilometers [233 miles] range;
- *al Kahar* - 1,000 kilometers [621 miles] range;
- *al Raid* - 1,000 kilometers [621 miles] range;

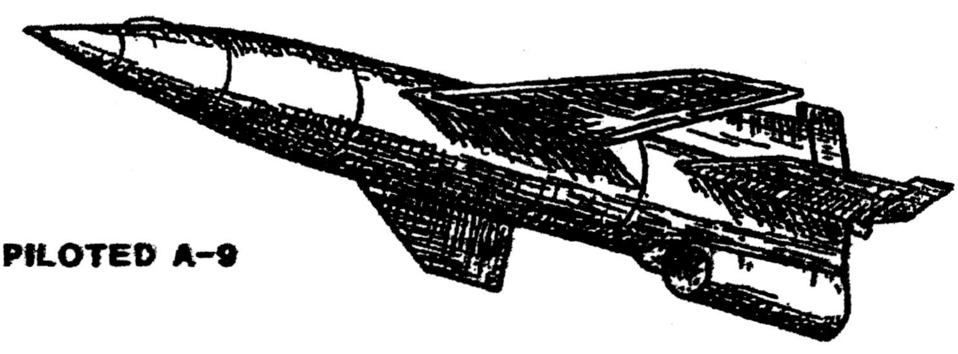

PILOTED A-9

A pen and ink drawing of *Wernher von Braun's* proposed winged, piloted *A-9.* Beneath the aft fuselage is a ram-jet auxiliary engine to help provide the *A-9* the needed range of 3,200 to 3,500 miles to New York City depending where in Europe the flying machine was launched.

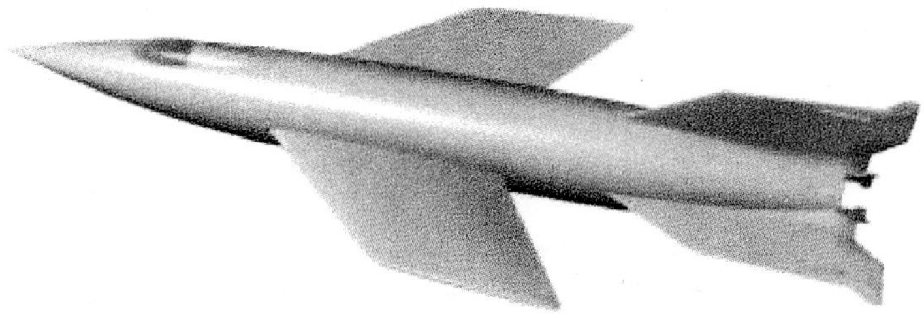

A nose port side view of *Wernher von Braun's* winged, piloted *A-9*. Digital image by *Josha Hildwine*.

Whether or not these allegations were true, the government of West Germany had no other choice than to rid itself of any association with *Sänger* and his assistants, to avoid the moral dilemma confronting them from the world community. The *Sänger* 'embarrassment' simply had to addressed...and quickly. The Bonn Government's decision was to put distance between itself and *Sänger* by firing him as the director of the West German Institute. It appears this decision was forced, because at the time of *Sänger's* firing, West Germany was making demands to major western military powers to give them a stronger presence and influence within NATO and to gain more access to information on military nuclear weaponry in general. The *Sänger* scandal was featured world-wide in headlines. West German morality was now being questioned at the very time they were seeking more influence in world-wide political and military affairs. Thus, *Eugen Sänger* and his team of rocket scientists became "politically expendable."

After being dismissed from his post as head of West Germany's Research Institute for the Physics of Jet Propulsion, *Eugen Sänger* and his wife emigrated to Egypt, where he became chief of the new

One of *Wernher von Braun's* former *A-4s* (*V2*) post war in the United States seen on its launch pad, 1946.

A successful vertical launch of a former *A-4* (*V2*) rocket missile in the United States. About 1946.

There was a fundamental difference between *Eugen Sänger* and *Wernher von Bruan* on how their *Amerika Bombers* should be launched. *Sänger* believed a horizontally launched, piloted, winged, reusable, manned bi-fuel liquid rocket was the most economical way of delivering a bomb to America. *Wernher von Braun* disagreed. He thoroughly believed in a vertically launched rocket from his beginning days. *Rudolf Nebel* (left) and *Wernher von Braun* (right), about 1930.

***Wernher von Braun* ...late in life.**

Ignition, lift-off, and the beginning of a vertical launch for the United States' *Space Shuttle*...following the principles *Wernher von Braun* established for launching his rockets back in the early 1930s.

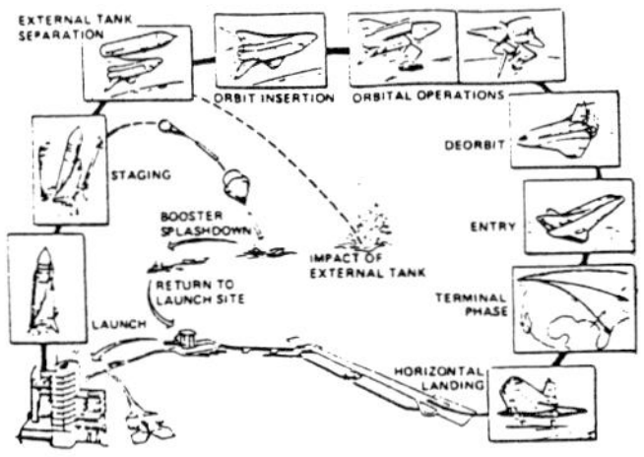

Typical Mission Profile

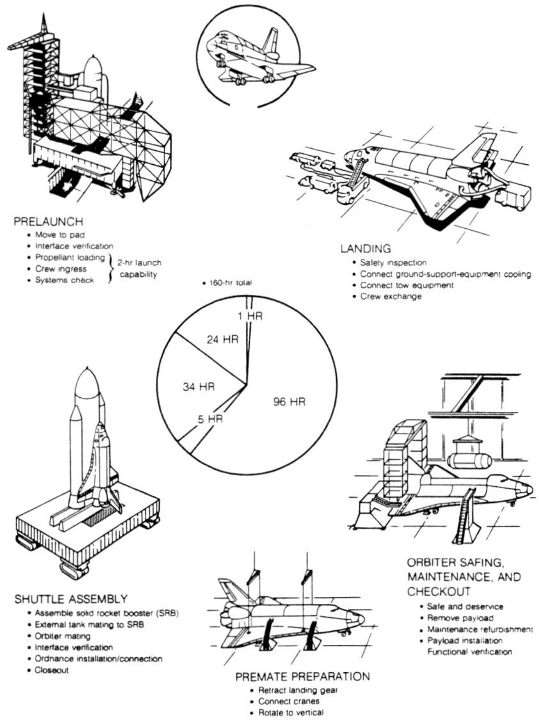

The Space Shuttle requires two weeks of preparation to turn it around for its next launch.

Left: The idea of horizontal launching proposed by *Eugen Sänger* over 55 years ago in *Nazi* Germany is barely alive in 2001. *George Retseck* and *Richard Post* have proposed in 2000 what they call the "inductrack launcher." These men claim their approach as illustrated could provide the initial boost for future spacecraft without all the manpower, infrastructure, and expense of the current vertical launch system. For example, the spacecraft could be mounted on a levitating "launch cradle" that would glide up a sloping one-kilometer-long track. Unimpeded by wheel friction, which *Sänger's* sled would have experienced, the cradle could accelerate the spacecraft to 590 miles/hour [950 kilometers/hour]. The spacecraft's rocket engines would then ignite and propel it into orbit...just as *Eugen Sänger* proposed in the mid 1940s. Courtesy: *George Retseck* and *Richard F. Post*, "*Maglev*: *A New Approach*," *Scientific American*, page 82, January 2000.

***Eugen Sänger's* winged reusable orbital rocket bomber poised on its 1.86 mile long monorail launch moments prior to ignition. Its mass was 100-ton. NASA's Dryden Research Center's *Al Bowers* calculates that it would have taken up to 30 *A-4* (*V2*) rocket engines to push it up to 1,200 miles/hour within the 1.86 mile- length of the monorail. Digital image by *Mario Merino*.**

rocket facility near Cairo. *Sänger* was reported to be earning $500,000 annually as full-time chief of *Nasser's* rocket research facility. Two other members of *Sänger's* former staff at *FPS* accompanied *Eugen* and *Irene* to Cairo. Critics alleged that *Eugen Sänger*, whether he liked it or not, would be forever marked as the man who was making possible *Nasser's* dream of vanquishing Israel.

Sänger Dead of a Heart Attack at Age 58

Eugen Sänger also found employment as a lecturer at the University of Berlin. He did not live long enough to see the realization of his dream of the "*Silverbird*," however. Three years before his death, on April 12th 1961, he learned that Russian *Yuri Gagarin* had first succeeded in a manned orbital flight around the Earth. Then, more than 5 years after *Eugen Sänger's* death, American *Neil Armstrong*

***Al Bowers*. Spring 2000.**

From the ground, *Sänger's* winged reusable orbital rocket bomber appears only a matter feet above its 1.86 mile-long monorail launch as it now fires up its internal 100-ton thrust liquid rocket engine. Digital image by *Mario Merino*.

***Eugen Sänger's* winged reusable orbital rocket bomber is climbing and showing us its flat fuselage underside which it will utilize as its skips its way around the earth propelling itself off the dense atmosphere. Digital image by *Mario Merino*.**

was the first man to set foot on the surface of Moon on July 21st 1969 thanks to *Wernher von Braun's Saturn 5* rocket whose first stage had generated 7.5 millions pounds of thrust.

Following *Eugen Sänger's* death, his wife continued to be active in many rocketry organizations. For nine years, *Irene Bredt-Sänger* was the scientific head of the German Association for Rocket Technology and Astronautics, Incorporated. She was co-editor for various aerospace journals and was received as a member of the International Astronautical Academy. In 1970, she was awarded the *Hermann Oberth* gold medal. However, critics noted that after her husband's death in 1964, her published works gradually changed in tone from detailed technical research to a more general historical, popular, and visionary character. This change, said critics, appeared to be in the nature of an effort to establish her husband's proper place in the history of the development of rocketry and space flight...stemming from her great personal admiration and devotion

What *Eugen Sänger's* winged reusable orbital rocket bomber might have looked like as seen moments after leaving its 1.86 mile long monorail launch, its 100-ton thrust liquid rocket engine leaving smoke and gas trailing behind. Actually, this is a photo of a *Messerschmitt Me 163, Walter HWK-509* powered interceptor as seen from its port side.

What *Eugen Sänger's* winged rocket bomber might have looked like moments after leaving its 1.86 mile long monorail launch. Its liquid rocket engine would have left a visible trail as has this *Messerschmitt Me 163, Walter HWK-509* powered interceptor seen in the photo.

The *Sänger* wing reusable orbital rocket bomber reaching for orbital flight altitude with all three of its rocket engines burning at full thrust. During this period, the winged orbital bomber would have consumed 90-ton of liquid oxygen and fuel in reaching 80 to 100 miles altitude in about five minutes. Digital image by *Mario Merino*.

to him. *Irene Bredt-Sänger* felt strongly that she and her husband were shunted aside by political and professional rivalries during the 1930s and were prevented from playing a more prominent role in German rocket developments. She died October 20th 1983, at the age of 72 years, but not before the West German government finally honored her husband's rocket work by naming their proposed reusable space vehicle "*die Sänger*."

The *VfR:* German Society for Space Travel

In 1927 in Germany, an organization was formed known as the "*Verein für Raumschiffahrt*" or *VfR* and meaning Society for Space Travel with *Johannes Winkler* as its first president. *VfR* offered mutual encouragement to its members, pooled its small resources, and conducted fund-raising to finance rocket experimentation. They published a small monthly magazine called "*Die Rakete*" edited by *Winkler* and among *VfR's* membership was *Willy Ley*, *Max Valier*, *Walter Hohmann*, *Guido von Pirquet* from Austria, *Hermann Oberth*, and others. *VfR* experienced various triumphs and failures, however, and public donations pretty much dried up in the world-wide economic depression of the early 1930s. *VfR's* final end came in 1933, when the incoming *Nazi* government of *Adolf Hitler* declared

Right: *Irene Bredt-Sänger* and *Eugen Sänger* seen in postwar France in the late 1940s.

The *Nord 1500 Griffon II*. The *Griffon* (originally known as the *Guépard*) was an experimental aircraft designed to flight test an airframe equipped with a combination turbojet-ramjet propulsion engine. The first prototype *Griffon I* made its first flight on September 20th 1955, powered by a standard *SNECMA Atar 101F* turbojet engine with afterburner. After its initial flight testing was completed the *Griffon I* was modified to complete the tests for which it was built. In this case the 7,716 pound static thrust French *Atar 101 E3* which was mounted inside a 4 1/2 foot diameter integral ramjet of the *SNECMA/Nord* design just forward of the ramjet's burners. The re-engined *Griffon I* was designated the *1500-02 Griffon II*. During its testing it exceeded *Mach 1* in level flight on May 17th 1957 with its ramjet ignited although not utilizing the full potential of the combination power plant. France claimed that May 17th was the first time a ramjet-powered aircraft exceeded the speed of sound anywhere in the world.

that rockets were to be used only in the national defense of Germany.

There was a bit more, politically, involved with this decision, however. Some of *VfR's* members, like *Willy Ley,* were actually strong resisters to *Nazism. Oberth* was an off-again, on-again pacifist. During the time of *Nazism*, *Oberth's* name and work were suppressed. German publications from that era are careful never to mention *Oberth's* research at all, though all of the features of the *V2* rocket missile and similar Peenemünde Experimental Rocket Center developments, for instance, were heavily based on his suggestions and investigations. It is safe to say that without *Hermann Oberth*, the *V2* rocket missile would not have been conceived. The *VfR* membership included many resisters and pacifists, which is the real reason why this space-flight society was broken up. Although state monies could be provided for such organizations in Germany, any such organization receiving the monies had to be state-controlled. Before this, rocket researchers had enjoyed a free exchange of ideas worldwide. However, after 1933, all rocket research in Germany was conducted and controlled by the State and a veil of tight security was wrapped around it.

A poor quality photo of *Eugen Sänger* and his wife *Irene Bredt-Sänger* seen at their drafting table in postwar France in the late 1940s.

The *Nord 1500 Griffon II* in flight as seen from its rear starboard side. Courtesy: *Henry Matthews*.

Sänger the Austrian

Eugen Sänger was born in the village of Pressnitz (Bohemia) on September 22nd 1905. He completed his doctoral studies in aeronautical engineering in July, 1929, with a thesis tilted "The Statics of Multi-spar Truss Wings with Parallel Webs, Cantilevered and Half-supported, Directly and Indirectly Loaded." In the Summer of 1957, he suffered his first heart attack. He suffered a 2nd heart attack while lecturing at West Berlin's Technical University on February 10th 1964, which proved fatal. Today, almost forty years after his death, he is still considered one of the most brilliant and inspirational theoretical physicists who worked in field of rockets and space vehicles. Although *Sänger* is perhaps best known for his World War II global-range rocket-powered bomber concept, postwar he also investigated the theoretical use of photon and light-powered rockets. *Eugen Sänger's* greatest contribution to the field of space rockets and space vehicles (aeronautics), may have been that of a dreamer-visionary who conceived ideas others brought to fruition.

After his death in February, 1964, critics and admirers pretty much agreed that he and his wife *Irene Bredt-Sänger's* study of a global bomber, entitled "Rocket Propulsion of Long Range Bombers", written for the German *Luftwaffe*, became one of the best known works in space literature after the war. It was an extension of ideas for a commercial space transport that *Sänger* had talked about in the early 1930s. Their orbital bomber study of the early 1940s envisioned a vehicle launched by a rocket propelled sled from its 1.8 mile long track and then climbing at 30 degrees to 5,100 feet altitude, where its 100-ton bi-fuel liquid rocket engine would ignite and burn for approximately 8 minutes. During this 'burn,' the entire 90-ton load of liquid oxygen and diesel fuel propellants would be consumed and the bomber would reach a speed of 13,596 miles per hour...all without the vertical launching complexes costing billions of dollars that are in use today.

His life-long friend , *Dr. Walter Dornberger*, claimed that *Sänger* was generally better performing mathematical calculations than doing the practical side of rocket development work. "He was always testing and trying, but never was able to carry projects

A dramatic ground level nose view of the *Nord 1500 Griffon II* featuring its gapping air intake for its *Sänger/Bredt*-combination *SNECMA Atar 101 E3* turbojet-ramjet. Also featured is the *Griffon's* fixed forward stabilizer which appears to give the machine a tail-first layout. Courtesy: *Henry Matthews*.

The *SFECMAS/Nord 1500 "Griffon II."* Courtesy: *Henry Matthews*.

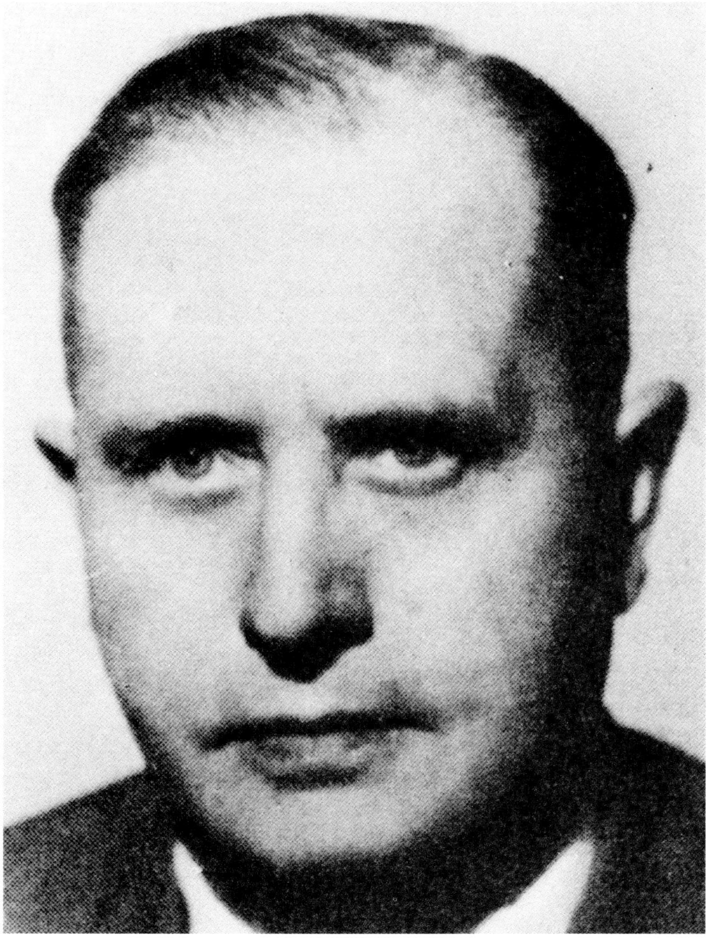

***Dr.-Ing. Hermann Oestrich* (1903-1973), former director of turbojet engine development at *BMW*. His efforts saw the successful completion of the *BMW 003A* turbojet engine. Postwar he was invited to continue his turbojet engine work and development in France.**

through," said *Dornberger*. "It appeared that whenever mishaps occurred, *Sänger* lost interest and went on to something else. Still, *Dornberger* said, "*Sänger* was very creative."

Eugen Sänger said he fell in love with the idea of space travel at about age 13 after reading *Kurd Lasswitz's* science fiction novel "*On Two Planets*" and he dreamed of doing something similar in his own life time. It was *Lasswitz* who introduced the term "*repulsor*" for the rocket engine and the concept of a space station. When *Sänger* enrolled in the Technical University of Graz, Austria, in 1923, he intended to major in civil engineering, focusing on construction of railroad and automobile bridges. However, in 1923, *Professor Hermann Oberth's* book "The Rocket Into Interplanetary Space" was published. *Oberth* was a German citizen living and teaching school in Romania. His book's three parts dealt with general prin-

***Professor Walter Georgii* in France. The former director of *DFS*-Ainring was invited by the French postwar to continue his research ideas and activities in France. After leaving France, *Professsor Georgii* for a time consulted for *Kurt Tank* in Argentina before returning to Germany. It was a happy time for *Georgii*, because he met his second wife in Argentina...his first having died.**

***Felix Kracht*, chief of design at *DFS*-Ainring (right), was invited to France postwar. He took with him his completed plans for the *Walter HWK-509* powered speed of sound research flying machine *DFS 346*. Others in the photo include *Ernst Udet* (left) and troop transport glider designer *Hans Jacobs* (middle).**

ciples of rocket operation and flight, high-altitude research proposals, and space flight. *Oberth* also had chosen liquid fuels for his rocket's operation because of their high exhaust velocity. *Sänger* said that *Oberth's* small scale best seller so thrilled and inspired him that he switched his major to aeronautical engineering from civil engineering.

Later, while still in college, *Sänger* came under the influence of the *Viennese School* of space research and its leader, *Max Valier,* who proposed propelling aircraft with rocket and jet engines. *Valier*, had written a book "*Die Vorstoss in den Weltenraum*" (The Plunge Into Interplanetary Space) which further popularized *Oberth's* earlier book. *Max Valier* later performed many spectacular experiments in several attempts to rocket-propel automobiles, sailplanes, and even iceboat racers. Many students and writers of rocket science feel that *Max Valier's* so-called sensational experiments added nothing to rocketry and only embarrassed *Valier's* colleagues by placing rocket science in a circus-like setting. Nevertheless, the space vehicle researchers at the *Viennese School* at this time – in the late 1920s and early 1930s – can be considered as the 2nd generation of space pioneers. They differed from the 1st generation in that they

The vertical lift off and landing of the "*Coleopter*" which *Count Helmut von Zborowski* designed and built in postwar France.

Former chief of aircraft design development at the *RLM* for *Hermann Göring*, especially the *Amerkia Bomber* projects, *Oberst Siegfried Knemeyer* for a time was working in France postwar. Then he received an offer from the United States which he couldn't refuse. *Knemeyer* was the unofficial boss of the *RLM* between 1944-1945.

Zborowski's "coleopter" shown in its travel/trailer in France.

tended to romanticize the possibilities of space flight without admitting the work and technology required to make it all happen.

Sänger also had been influenced by another member of the *Viennese School*, *Professor Franz von Hoefft,* who was researching the possibilities of achieving necessary velocities for space flight by the combustion of liquid fuel hydrocarbons. He also had proposed a rocket powered lifting body-like vehicle capable of taking off and landing like a hydroplane, with only the upper stage having orbital capabilities. *Sänger,* caught up in the romantic possibilities of space travel, realized, that an organization capable of directing millions of man hours would be required to make the dream of space travel a reality. 'Lone wolves' like *Oberth*, *Valier*, and *von Hoefft* could not hope to succeed by doing 'hobby' and 'casual' work – merely by attaching rocket engines to automobiles and sailplanes. The necessary organizational structure, human resources and funding could be easily obtained, *Sänger* believed. The technically difficult part of the whole concept was propelling the vehicle away from earth and into space. *Sänger* decided he'd devote his working life to perfecting the space rocket's propulsive system.

Based on ideas presented by *Hermann Oberth* in his 1923 book, *Max Valier* in 1925, and *William Ley* in 1926, it appeared that a space vehicle would take on a purely ballistic form, propelled by a multi- stage, liquid propellant rocket engine. *Valier*, for example, said:

Zborowski's "coleopter" is seen being lifted up and out of its travel/ trailer in preparation for setting the flying machine on the tarmac for a test flight in France.

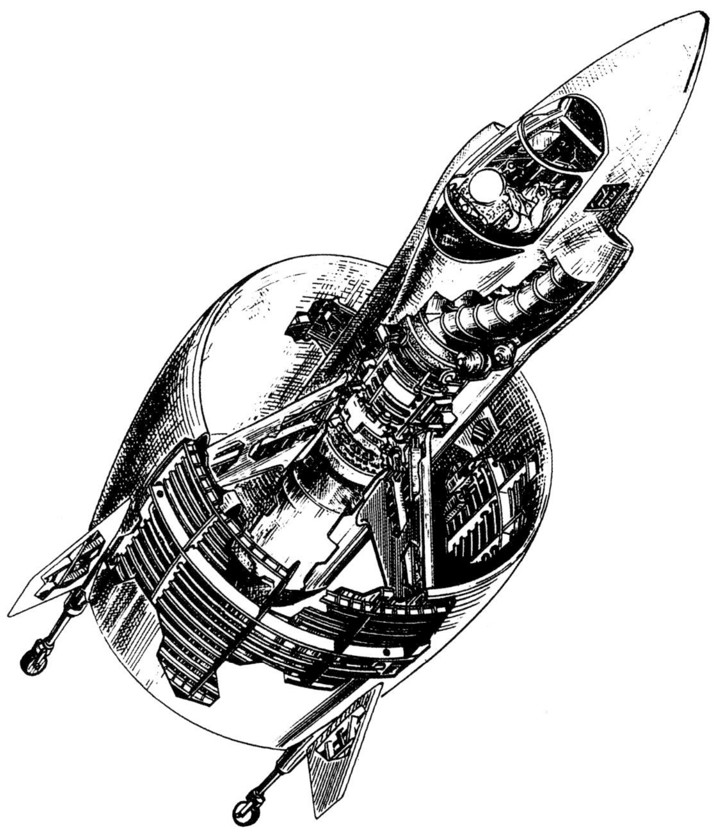

A pen and ink cut away view of *von Zborowski's* "Coleopter" he designed, built, and test flew (it crashed attempting to land on its first flight and never was repaired) in France postwar.

A steam rocket test stand constructed at Echterdingen-Stuttgart for *Eugen Sänger's* Research Institute for the Physics of Jet Propulsion (FPS).

***Dr.-Ing. Eugen Sänger* (left) and *Dr.-Ing. Walter Dornberger* (right) with a scale model of *Sänger's* LO^2/LH^2 rocket engine. This scale model was built by *Bell Aircraft* according to *Sänger's* patents. About 1955. *Dornberger* regretted that he didn't use greater persuasion in order to get his friend *Eugen* and his family to come to work at *Bell Aero-Space*. If he had, he believed, *Sänger* might have never agreed to consulting with *President Nasser* of Egypt and the professional disaster which followed.**

GERMANY PLANS 2,000 m.p.h. PLANE

10-YR. PROJECT, SAYS JET EXPERT

From LEONARD BERTIN,
Daily Telegraph Science Correspondent

COPENHAGEN, Monday.

The idea of a 200-ton airliner that will travel at three times the speed of sound, more than 2,000 m.p.h., is being studied in Germany's recently formed Research Institute for the Physics of Jet Propulsion. Dr Eugen Saenger, director of the institute, told me this here to-day.

The institute is financed jointly by the West German Federal Government, the State of Württemberg-Baden, and a number of commercial companies, which include aircraft firms like Heinkel, Messerschmitt, Dornier, Bosch and the Daimler-Benz company.

Dr. Saenger, who is one of the official German delegates to the International Astronautical Federation congress, which opened in the technical university here to-day, told me that the institute had a two-fold "command" from the German Minister of Tran[illegible]

The first task [illegible] investigate the economics of [illegible]e termed a "hot water rocket" that might give transport and passenger planes a powerful take-off boost at negligible cost. The second was the provision of a complete set of dimensional tables and data for ram jet engines for planes of all reasonable sizes and speeds

STEAM JET

The hot water rocket, known technically as the Ruth-Speicher principle, is basically more than 100 years old. Roughly speaking, the idea is to get a strong steel bottle, fill it [illegible] and stopper it, and then

This item appeared in the *Daily Telegram*, August 2nd 1955 reporting that *Dr. Eugan Sänger* of Germany's recently formed Research Institute for the Physics of Jet Propulsion, was planning a 2,000 mile/hour, 200-ton airliner.

"Consider, please, my idea of developing the spaceship from Junkers aircraft. I imagine such aircraft initially provided with Blicharski's wing propeller and to be completely hermetically sealed, with normal air pressure inside. Then there would be a way of developing an "intermediate type" between aircraft and spacecraft. Such vehicles would take off like aircraft. Thereafter at an altitude, where propeller engines fail to work, they would boost themselves with a single rocket impulse. Like the projectile of a long-range cannon, they would thereupon follow a very smooth ballistic trajectory over most of the flight path, where the characteristics of a free flight trajectory are only slightly influenced by the effect of the airfoils. During the descent path, from reentry altitudes of 12,000 to 10,000 meters, they would again proceed to the normal gliding flight of an aircraft. Their attainment of half orbital velocity, together with the lifting effect of the airfoils, would greatly increase their gliding path. Finally, they would land just like ordinary aircraft. I cannot depart from the concept of a spaceship with wings, especially not from the "intermediate type," a rocket plane which daily appears to me to have greater chances of realization. A rocket plane using gasoline as fuel and compressed air as oxidizer could operate with the oxygen of the atmosphere still in altitudes of 15 to 20 kilometers and attain burnout velocities of 1,400 to 1,500 meters per second, in my opinion. In this way one could eliminate the difficulties associated with liquid oxygen carried on board."

An artist's rendering of *Bell Aircraft's* bomber/missile (*Bomi*) proposal to the USAF in 1957. Its mission was to provide a piloted, maneuverable vehicle and associated equipment for conducting experiments in hypersonic and orbital flight altitudes. It was to gather research on controlled, maneuvering reentry from orbital flight. In addition, it was to demonstrate a pilot controlled maneuvering reentry with tangential recovery at a pre-selected landing site. *Bell's Bomi* origins can be traced back to *Eugen Sänger* and *Irene Bredt's* antipodal orbital bomber. In the early 1950s, *Bell Aircraft's Walter Dornberger* and *Krafft Ebhicke* began working on a proposal for a rocket-launched global-range bomber. This became the *Bomi. Walter Dornberger* invited *Sänger* to join him at *Bell Aircraft* to help them on the *Bomi* because *Bell* had taken advantage of *Sänger's* wartime research on his orbital bomber project. The *Sängers* declined *Dornberger's* offer.

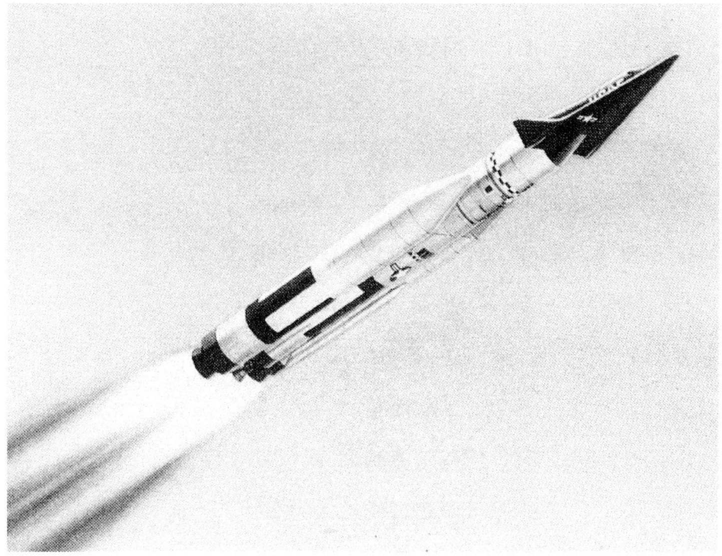

Launching the *Bell Dyna Soar* via a *Titan III* booster rocket toward orbit.

In Spring 1927, during a lecture given in Berlin, *Max Valier* continued to define his rocket- powered spaceship:

"At first the *Junkers G-23* will be equipped with two rocket engines in the airfoils. If jet propulsion proves to be successful within the first tests in flight, a true rocket plane with an auxiliary reciprocating engine retained for safety could represent the next transfer stage of development. First there will be four, and later on, six rocket engines in the wings. When sufficient experience has been gained, the pure rocket plane, with small wings and a pressure cabin suitable for intercontinental flights in the stratosphere, can be built. The final stage is to develop a rocket ship which will ascend vertically from a launching tower, and which will be developed into the spaceship at last."

The two solid rocket boosters attached to the *Titan III* are seen being dropped after their fuel was spent.

The *Dyna Soar* and the *Titan III's* adaptor stage have separated from the *Titan III's* liquid core, which will now provide the final push to place *Walter Dornberger's Dyna Soar* into orbit.

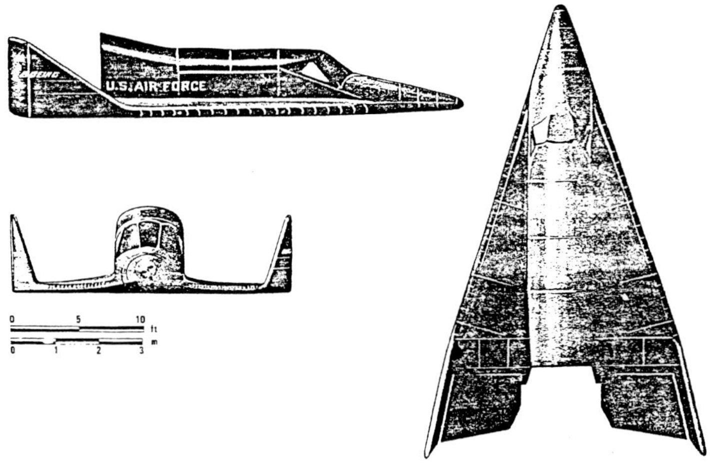

A poor quality pen and ink drawing of *Boeing's X-20 "Dyna-Soar"* project abandoned by the USAF in 1963 before flight trails could begin. Conceived in 1957, it would have been America's first manned spacecraft after the *X-15*.This machine was designated *X-20*, was based on *Eugen Sänger's* orbital rocket bomber of *World War Two* which could skip around the earth utilizing the upper atmosphere following the principles laid down twenty years before by *Eugen Sänger*. With *"Dyna-Soar," Boeing* and others had hoped it would be an important part in developing aerodynamic and structural techniques for new-generation spacecraft capable of military maneuvering from orbit. As an orbital bomber it would have carried "stand off" nulcear weapons into orbit. Only one prototype was constructed. Its length was 35 feet [10.7 meters]; wing span 20 feet [6.09 meters]; and landing would have been accomplished via wire-brush landing skis.

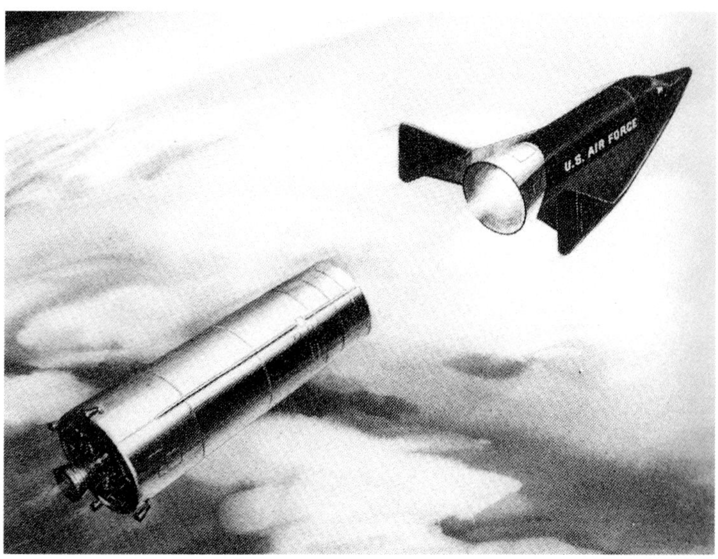

The *Dyna Soar* is now free from the *Titan III* and is on its own. Reentry would have required the pilot to slow the flying machine around and firing its rocket to serve as a retro brake.

Boeing's final mockup of their single-seat *"Dyna Soar"* features a rugged structure as viewed nose on.

***Bell Aircraft* and *Walter Dornberger's* "aerospaceplane" which was intended to reach any contract on earth in one hour or less. The continent-spanning craft would be launched by a mother aircraft "piggy back" style and after the launch would return to its home base.**

Reentry of *Bell Aircraft's* orbital rocket bomber/transport. The heat of reentry has caused the wing's leading edges to glow due to the enormous heat generated.

Max Valier died in May, 1930. The liquid oxygen/gasoline rocket engine which he had placed in a specially designed vehicle exploded one evening at the empty factory building in Berlin where he was working on it. All alone in the abandoned factory, a metal splinter from the blast pierced his aorta and *Valier* bled to death with no one knowing about it.

It appears *Eugen Sänger* also was highly influenced by *Dr.-Ing. Franz von Hoefft,* one of the leaders of the *Viennese School*. In a lecture in Vienna in February 1928, *von Hoefft* theorized the form a spaceship might take.

"I contemplated every possibility, from the exhaust of compressed air to that of either atoms and electrons accelerated by zero-point energy of ether, or by the energy of nuclear fission, until in 1924, I read Oberth's book...and realized the possibility of achieving the necessary cosmic velocities by using as fuel liquid hydrocarbons presently available."

Von Hoefft's concept of a spaceship was different from those of either *Oberth,* or *Valier*. Instead *von Hoefft* proposed a spaceship shaped like a winged lifting body. In 1927, *von Hoefft* had the idea of testing a model rocket in the wind tunnel at the Aerodynamic Institute of the *Technical Hochschule* in Vienna. This model was based on a design presented by *Guido von Pirquet*. *Pirquet* said later that his model produced satisfactory results in the wind tunnel testing, but had no immediate practical application at the time. About this time, *Pirquet* said that he had heard of a young scientific assistant at the Institute who was interested in rocket problems. This person was *Eugen Sänger*. In March, 1927, *Sänger* applied for membership in the *Wissenschaftliche Gesellschaft für*

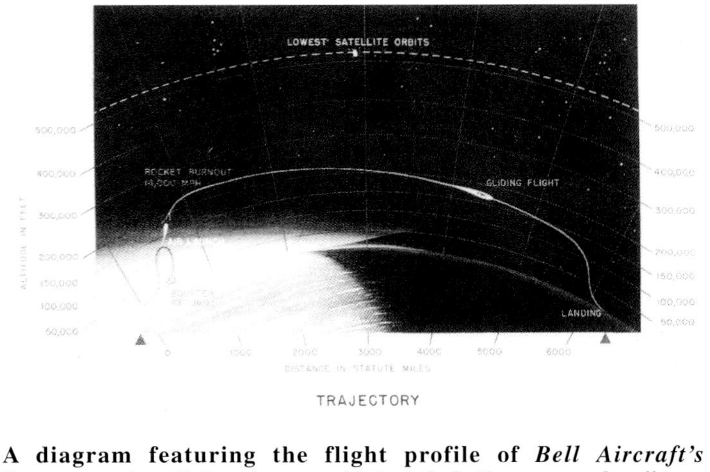

A diagram featuring the flight profile of *Bell Aircraft's* "aerospaceplane." *Dornberger* said that their "aerospaceplane" was based on his friend *Eugen Sänger's* winged reusable orbital rocket bomber project of *World War Two*.

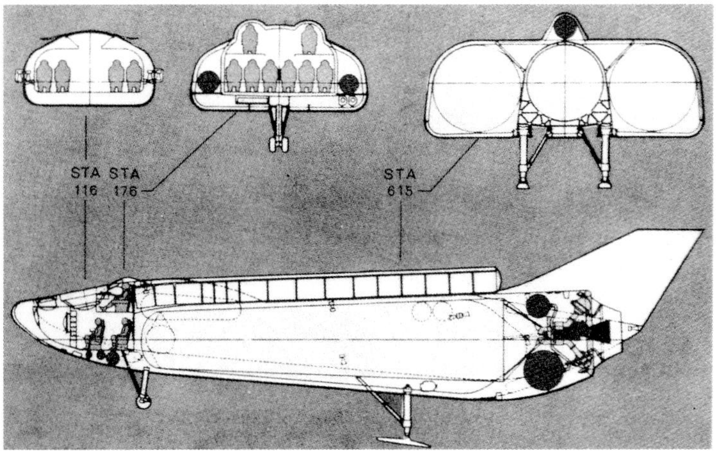

A *Lockheed* concept for a cargo/passenger carrying stage for a proposed flyable "aerospaceplane." As with the *Sänger* orbital rocket bomber, *Lockheed* aero-engineers had to reserve most of the flying machine for its fuel tank. The space allocated in the *Lockheed* proposal shows the enormous quantities of fuel required for space flight.

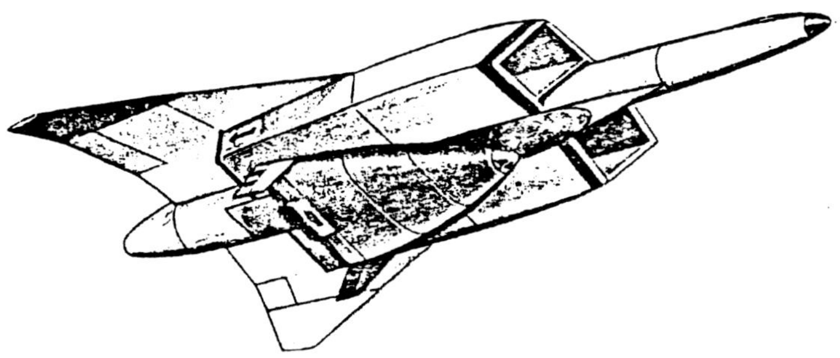

A proposed French/West German space transporter by *Nord*, *Nord-Aviation*, and *SNECMA*. This winged orbital flying machine would have been carried beneath a large air-breathing booster which would have been able to take off and land conventionally on a runway. Similar in several respects to *Eugen Sanger's* winged orbital rocket bomber concept of the early 1940s.

Höhenforschung (Scientific Society for High Altitude Research). In his letter to *von Hoefft*, the Society's secretary, *Sänger* volunteered to be of assistance to their upcoming spaceship shape testing in the *Wissenschaftliche* wind tunnels. *Sänger* did not participate in the wind tunnel testing of *von Hoefft's* scale model spaceship. He was very involved in his Ph.D. studies and duties as a scientific assistant at the college. In addition, *Sänger* – to those who knew him – said many times that the most difficult stage in the realization of space flight was not in the shape of the hull nor its aerodynamics, but rather centered on its propulsion device. It was this problem which *Sänger* seemed most interested in. He devoted most of his energies throughout his life to this problem.

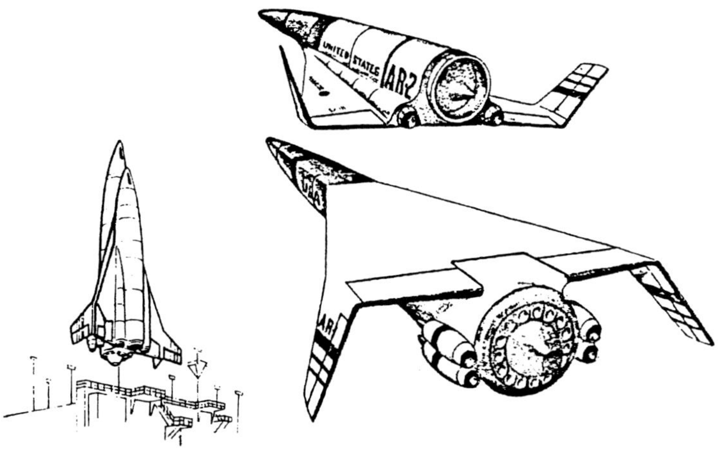

Vertical takeoff of the two-stage "*Astrorocket*" (*AR-2*) proposed by the *Glenn L. Martin Aircraft Company* of Baltimore, Maryland (left) in the early 1960s. The "mother carrier/booster" would have been powered by a large bi-liquid rocket engine plus four large thrust turbojet engines attached to the aft outer fuselage. The "mother carrier/booster" carries the top stage which is powered by a bi-fuel liquid rocket engine during takeoff to orbit. Two additional large turbojet engines would have added additional thrust. The *AR-2's* stage 2 would have left orbit via a glide-return and landing. Stage 2 of the *AR-2* would have followed similar principles returning home as laid down by *Eugen Sänger.*

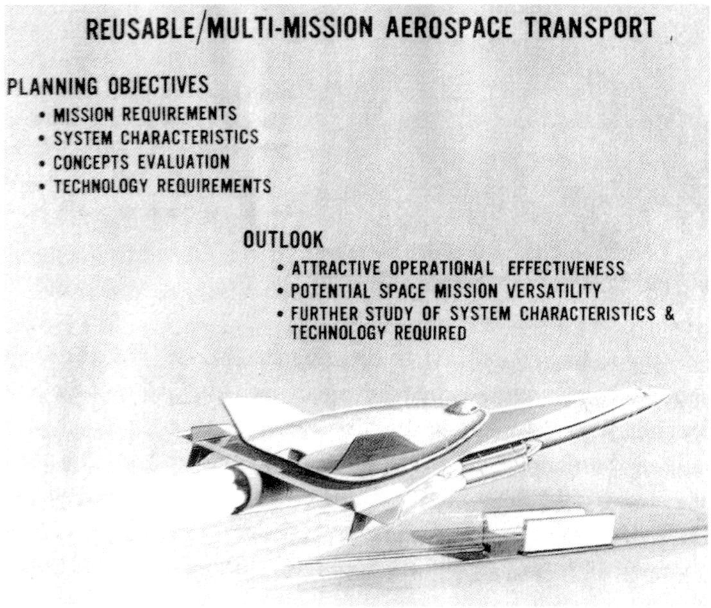

An aerospace transport proposed by *NASA*. On the lower right can be seen the aerospaceplane shortly after leaving its "mother carrier/booster's cradle. The benefits cited include reusable, multi-mission, and cost effective...the same argument *Eugen Sänger* stated in his proposed to the *RLM*, *DVL*, and others some fifteen years earlier in *Nazi* Germany.

When the *Nazi* government nationalized all private and public rocket research in 1933, *Sänger* discovered there was very little research data and information written down on rocket burn time, performance, and fuel consumption from the former *VfR*. Besides this, all experimentation had now become a State secret. Yet, in reality, these 1st generation German space pioneers operated in a "hit or miss" world where experiments that worked.. worked and

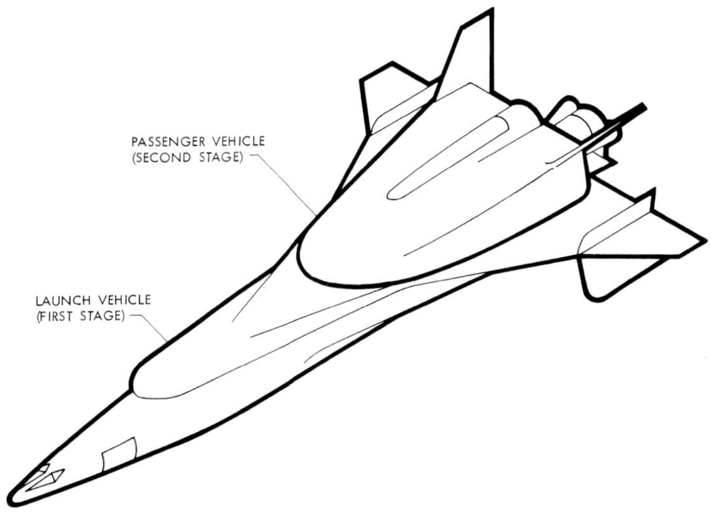

A pen and ink design proposal by *Lockheed* for a winged, reusable, wheeled landing booster (1st stage) for launching aerospace planes with a passenger carrier (2nd stage).

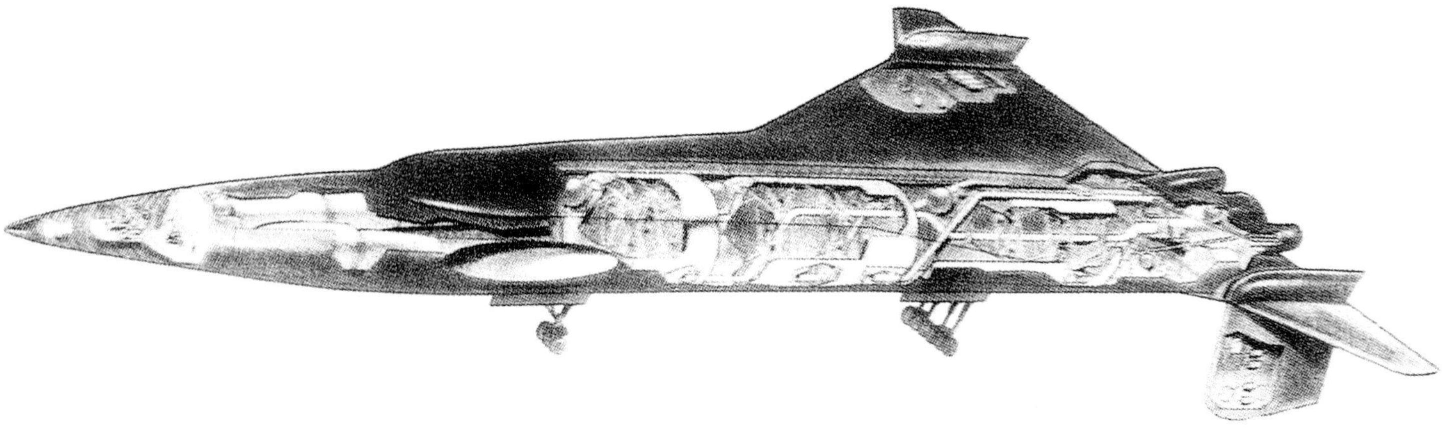

Pen and ink illustrations by *Lockheed Corporation* featuring their proposed winged, reusable, wheel landing recoverable booster for launching "aerospaceplanes." The orbiting flying machine is mated/carried on the "aerospaceplanes'" upper back.

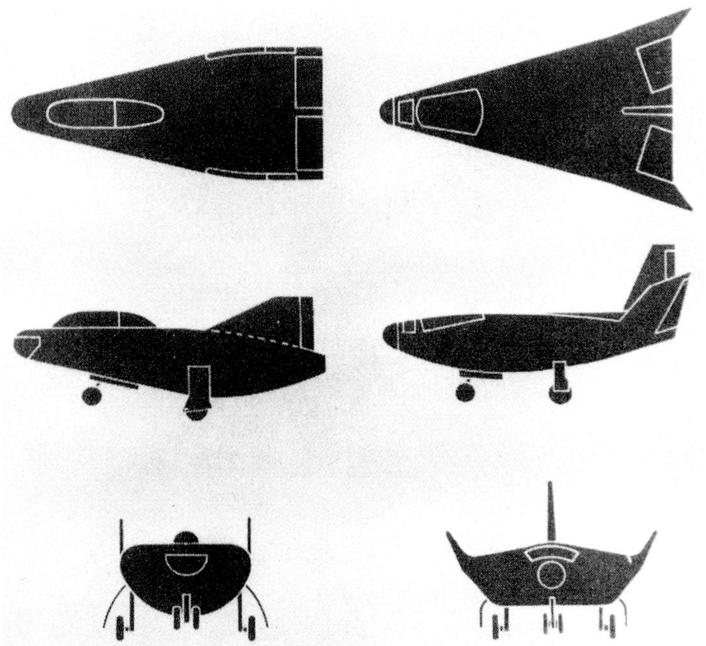

"Winged craft" or "lifting body" design research began in the mid to late 1950s in the United States. One of the first serious efforts for a flyable shape for reentry from orbit was started by a team headed by *Dr. Alfred Eggers* of *NCAC*...predecessor to *NASA*. For five years *Eggers* and his team tested a large number of lifting body space craft shapes in the *Ames Research Center's* wind tunnels. *Eggers* designated these lifting bodies his *"M-series"* and the wooden *M2-F2*, built for *NASA* by *Northrop Corporation,* was finished in 1965. Fourteen subsonic flights had been accomplished with the *M2-F-2* through 1967 being dropped from a *Boeing B-52* turbojet powered bomber. *NASA* gave *Northrop* an additional contract to construct an improved version out of aluminum. It was known as the *HL-10* (*HL* for horizontal landing) and differed in considerably in configuration from the *M2-F2* in overall contour, cockpit design, and flight controls. It also carried three vertical stabilizers. The *Northrop Corporation's M2-F2* (left) and the *HL-10* (right).

The *Glenn L. Martin Aircraft Company's* unmanned reentry vehicle *SV-5D* which had been flight tested atop an *Atlas* booster rocket. *Martin* was requested to build a manned and rocket-powered (a *Reaction Motors XLR-11* rocket motor) version of the *SV-5D* that came to be known as the *Martin X-24-A.*

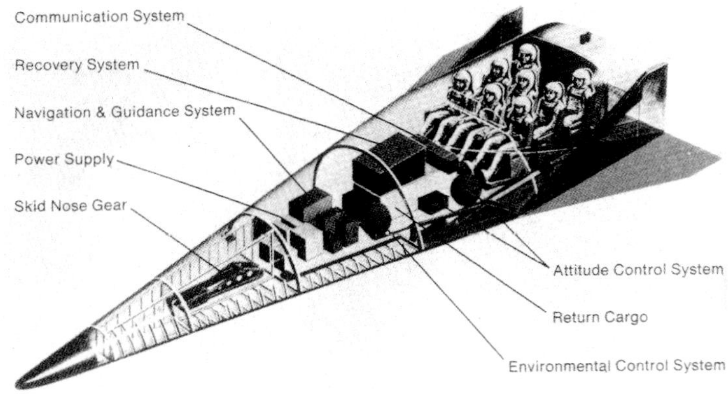

Northrop Corporation's **concept for a** ***Sänger*****-like lifting type flying machine capable of shuttling crews to and from orbiting space stations, as well as other missions, such as space rescue.**

The ***Glenn L. Martin Aircraft Company's X-24-A*** **as seen directly from the rear. The four nozzles of its** ***Reaction Motors' XLR-11*** **rocket motor...the same motor which powered** ***Bell Aircraft's X-1*** **through the sound barrier by** ***Chuck Yeager*** **on October 14th 1947.**

those that didn't...were ignored and forgotten. Furthermore, these early rocket experimenters often kept their successes and failures to themselves, seldom documenting, or publishing their results. German higher academia, research institutes, and heavy industry, too, had not contributed much to the new science of rocket propulsion.

In 1933, *Sänger* was involved in rocket research in his native Austria with little or no financial aid. He believed that high altitude flight could benefit from the use of rocket power, that is — replacing the internal combustion aircraft engine with rockets. *Sänger* continued with his research into constant pressure combustion engines, and working on a report to be submitted to the Austrian government, in order to request funding and support for his work. But the report didn't result in any public funding. The Austrian government turned down *Sänger's* request for public funding, saying that the use of liquid hydrocarbons and liquid oxygen was incapable of practical realization due to the excess danger of explosion. With this firm rejection and no further possibilities for government financial support, *Sänger* was forced to seek research grants and/or employment outside Austria. Throughout 1934, *Sänger* had sent applications for employment at several German research institutes across the Austrian border in Germany. Nothing positive happened with these job inquiries so *Sänger* continued on with his private research work. His interests now were getting him involved in the properties of various vaporized fuels and ways to cool constant combustion engines. He later wrote of his discovery:

> "Under continuous ignition, a mixture of atomized liquified oxygen (lox) and atomized diesel fuel burns very much like gaseous oxygen. Oxygen mist seems to ignite only after complete evaporation. The mixture of LOX and frozen fuel droplets does not detonate, but burns stable and quite rapidly."

His theory now proven to his satisfaction, *Sänger* applied for 12 different patents. He received patents in Germany, Italy, France, and England. His patent requests sent to the United States Patent Office were not granted and no explanations for non-issuance were given.

The ***Reaction Motors*****-powered** ***Glenn L. Martin Aircraft Company's X-24-A*** **lifting body vehicle as seen from its rear starboard side. This program was also known as** ***PILOT*** **(for Piloted Low Speed Test). Other USAF names used in this test program included:** ***START*** **(Spacecraft Technology and Advanced Reentry Tests);** ***ASSET*** **(Aerothermodynamicelastic Structural Systems Environmental Tests);** ***PRIME*** **(Precision, Recovery, Including Maneuvering Entry).**

Sänger's Research

Into Fighter Aircraft Assisted Take-off Liquid Rocket Units
In January, 1936, one month before he was hired by *DVL*-Berlin/Adlershof, *Sänger* published the results of earlier research studies he had made on liquid rocket units designed for take-off assistance for aircraft. It is not clear why *Sänger* researched this form of rocket use at this time, but it appears he was interested in helping fighter/interceptor aircraft climb rapidly to altitude in order to intercept bombers. A fighter aircraft of this period weighing about 2-ton [4,000 pounds] and capable of a level forward speed of 320 miles/hour, could climb to 20,000 feet altitude from the ground in about 8 minutes. *Sänger* claimed that he could reduce a fighter aircraft's 8 minute climb time to only 1 1/2 minutes. He would accomplish this by installing in the aircraft's fuselage three spherical tanks: liquid oxygen, fuel oil, and nitrogen gas needed to pressurize the fuel and liquid oxygen lines. Total installed weight was 950 pounds. *Sänger's* liquid rocket motor would burn for 90 seconds and deliver one ton [2,000 pounds] of thrust, with a jet velocity of about 8,000 feet per second. The empty weight of the rocket motor and its tanks was 205 pounds. The fuel weighed 745 pounds. *Sänger* stated in his article that he had already achieved jet velocities of more than 8,000 feet per second from his laboratory models. No prototype *Sänger* liquid rocket assist take-off unit is known to have been built and tested. It appears that the *RLM* was more interested in the externally mounted and 'dropable' versions from *HWK* than the permanently installed unit *Sänger* was proposing.

Research Into Ramjet Powered Fighter Aircraft At *DFS*
In the early 1940s, *Eugen Sänger* had discussions with senior staff officers from the *RLM* regarding the use of the simple ramjet engine for powering fighter aircraft. *Sänger*, gave few clues regarding the nature of these discussions, however, it appears the *RLM* was seeking fighter aircraft powered by ramjet engines. The Germans preferred to call the ramjet the "*Lorin Duct*" although some German researchers at the time like *Klaus Oswatitch* preferred the name "*Trommsdorff Drive*". The ramjet is among the simplest of engines. It produces the pressure increase within its combustion chamber due to the impact pressure at high velocities, so this type of engine uses no rotating or moving parts. However, ramjet engines do not

The "*Sanger* Projeckt" as proposed by MBB and viewed on the tarmac as seen from its port side rear. The reusable "Sänger" was named in honor of *Eugen Sänger* and would be carried aloft by the mother carrier/booster.

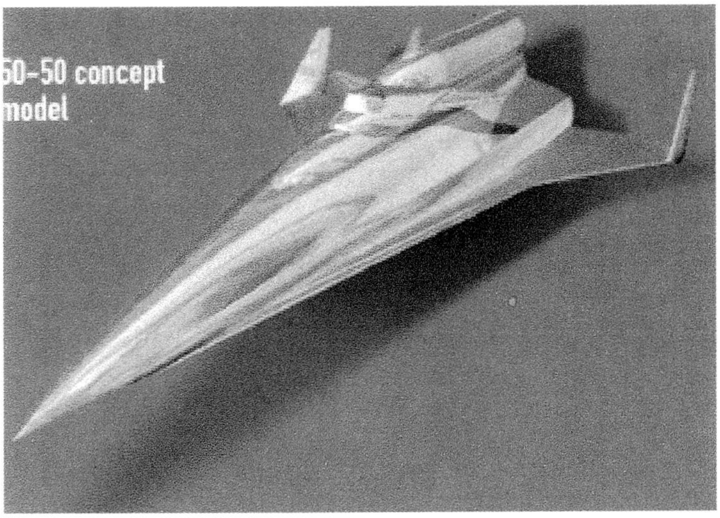

The *Mikoyan MiG 105 Spiral* and *50-50* concept. Soviet rocket designers had been fascinated by the concept of *Eugen Sänger* and his winged reusable orbital rocket bomber in April 1945 after sacking Berlin. In 1962 the *Mikoyan Design Bureau* proposed many variations of the space/plane concept. The "mother carrier/booster" would have been a hypersonic flying machine for takeoff and when reaching *Mach 5.5* would release the *MiG 105 Spiral* space plane on a piggyback booster to attain orbit. The hypersonic "mother carrier/booster" was never built, however, the *MiG Spiral* (*Lapot*) meaning flat iron, was.

New York Times Aug. 2, 1952

Germans Again Take Up Rockets, But in a Purely Theoretical Way

By JOHN HILLABY
Special to The New York Times.

INNSBRUCK, Austria, Aug. 1—German research into the theory of rocketry will be resumed officially this year in Stuttgart, where an institute is being organized by Dr. Eugen Sanger, a former German rocket expert. Until recently he was employed by the French at the Arsenal Astronautique near Paris.

This information was discussed informally tonight by members of the International Astronautical Federation who are assembling here for their fifth congress.

Dr. Sanger was not available for comment. He is expected at the congress. He and his wife, Dr. Irene Sanger Bredt, were employed on ram-jet research at the German rocket stations of Trauen and Ainring, in Bavaria, during World War II. Since then they have been considered to be among the best rocket physicists in Europe.

Theory to Be Sole Concern

According to letters received from Dr. Sanger and his wife by members of the Astronautical Federation, the new Stuttgart institution will be concerned solely with the theory of rocketry and its members will not be permitted to indulge in practical research. The financing will be restricted to about $32,000 a year, it is said.

Apart from this official interest in rocketry in Germany, the fifth congress has drawn delegates from commercial corporations and Government projects whose work is usually regarded as secret.

Delegates from the United States include Prof. Theodor von Karman of Pasadena, Calif.; Dr. Ernst Stuhlinger of the Redstone Arsenal of the United States Army at Huntsville, Ala.; Richard F. Comperz, United States Air Force Rocket Test Station, Edwards Base, Calif.; Dr. Hubertus Strughold of the Department of Medicine, Randolph Field, Tex., and executives of the Bell and Goodyear Aircraft Corporations and the American Rocket Company.

The new president of the international federation is Andrew G. Haley, a lawyer from Washington who is also president of the American Rocket Society.

Five More Societies Apply

The past president of the federation, Frederick C. Durant, a consulting engineer also of Washington, announced that five more societies had applied for membership in the federation. They are the Japanese, Egyptian, Brazilian, Spanish, and American Astronautical Societies.

He said that since the congress met last year "man has flown higher and faster than ever before." He considered the most meritorious feat to have been the flight of the Viking II sounding rocket, which rose to a height of 156 miles last May.

In his opinion, the space-flight problems of weightlessness, radiation hazard, aerodynamic heating and the mechanical reliability of propulsion systems "were becoming clearer."

However, he said, some problems such as the psychological and physiological effects of prolonged weightlessness "were unknown."

An article from the *New York Times* newspaper of June 2nd 1952 describing how West Germany was about to re-enter the field of rockets and missiles for the first time since its surrender to the Allies on May 8th 1945. This re-entry is described as only a theoretical undertaking and not the manufacturing of hardware. *Eugen Sänger* became the director of the Research Institute for the Physics of Jet Propulsion in the Summer of 1954.

produce any thrust at zero speed, so aircraft equipped with such propulsive systems must be accelerated to their required operational speed by means of a piston engine, piggyback towing, or by a liquid or solid rocket assist from a launching base or platform.

In 1912, the Frenchman *Réné Lorin* had taken out a patent on ramjet propulsion that was far ahead of its time, since this method of aircraft propulsion only becomes efficient at velocities near the speed of sound and above. He had made suggestions on ramjet propulsion beginning in 1908 in a series of articles published in the French aviation magazine *l'Aérophile*. *Lorin's* ideas were sort of an early jet engine, producing a very fast exhaust jet velocity of low mass verses the jet turbine engine, which produces a slow exhaust jet velocity of very high mass. *Eugen Sänger* was one of the first aerodynamists in Germany to become interested in the problems of propulsive ducts and while he had not developed his theory to the same degree as *Alexander Lippisch*, or *Otto Pabst* of *Focke-Wulf*, *Sänger* had acquired much more practical experience. He had gained this experience with actual ramjet models in flight on *DornierDo 17 Z-2* and *Dornier Do 217K V-3* aircraft and from ground tests on moving vehicles. However, *Sänger* conducted no known wind tunnel tests under controlled conditions so it is not possible to establish a correlation between his theory and his test results.

A nose-on starboard side view of the single seat orbital space-plane *MiG 105 "Spiral."*

4-Man German Team Built U. A. R. Rockets

By Oscar F. Gavrilovich
Herald Tribune News Service

When President Nasser recently climaxed the 10th anniversary of his overthrow of King Farouk by setting off four rockets, he proudly asserted that these were entirely of Egypt's own design.

He let the implication rest that some son of the people of the pharaohs had created the weapons. But the actual identity of their designer is well-known in West Germany, and in Israel, too. He is Eugen Saenger, former head of the Stuttgart Institute of Physics, the only space-rockets research institute in West Germany.

Dismissed from his Stuttgart post last year for spending too much time in Nasser's domain, Saenger was instantly rewarded with a handsomely paid job in Cairo, as chief of the new rocket-plant to be built in Egypt. With him went his wife, Irene Bredt-Saenger, who also is a physicist, and two other Stuttgart associates.

$500,000 in Salaries

For their services Nasser offered a yearly salary of about $500,000, half payable in advance. Saenger was also given Egyptian funds to buy in West Germany all the equipment and material needed to start the rocket project.

Saenger's name is well-known in advanced aeronautics circles. After the collapse of Germany, when the Allied powers were competing for the services of Hitler's top scientists, Saenger was interviewed in Upper Bavaria by American authorities; the United States, however, made no objection when Saenger and his wife accepted a French offer. Later the Saengers took up their researches at Stuttgart.

Saenger should by no means be classified only as a rocket designer; one of his primary interests has been the perfection of a "space glider." According to West German sources it is by use of formulae originally worked out by Saenger that the U. S. Air Force launched its Project Dyna-Soar and designed the fabulous X-15.

Designed Space Vehicle

Born 57 years ago in Austria, and graduated from Vienna's School of Technology in 1929, Saenger qualified one year later as an airplane pilot.

In 1932 he designed a model of a space vehicle to be set into orbit by rockets, then maneuvered by an astronaut. He peddled his invention and his mathematical data around government offices in Vienna, but was only laughed at.

However, in 1936, when Saenger offered his design to the German Luftwaffe, Hermann Goering saw value in the device. Goering created for Saenger the "test airplane center of Traunen" on the heath of Luneberg, then accessible only by parachute.

Saenger wrote for publication a 900-page manuscript covering his researches there. Among items he described were combat rockets capable of striking enemy territory from 1000 to 20,000 kilometers away.

In the meantime, World War II had begun and the Luftwaffe feared it might take 10 years to make Saenger's missiles operative. For security reasons, publication of his book was barred. Only 120 copies of significent excerpts from it were mimeo-

A newspaper column from the *Washington Post*, March 12th 1953, alleging that *Eugen Sänger* was earning $500,000 a year for assisting the Egyptians' design and development of offensive rocket missiles to be used against Israel.

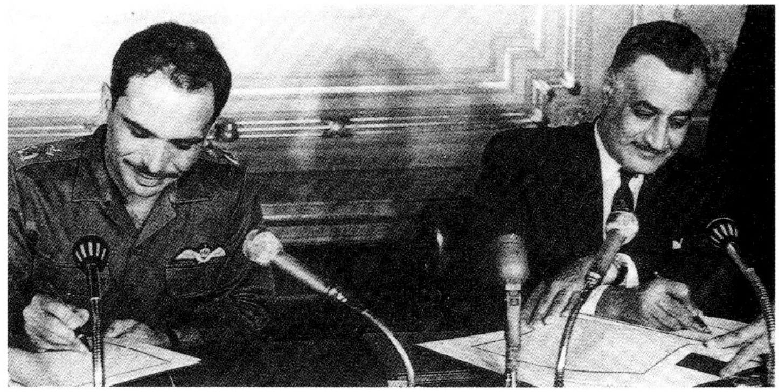

King Hussien of Jorden (left) and *Gamel Abdel Nasser* of Egypt (right) signing their 1967 mutual military assistance agreement in their on-going struggle to overcome the Israelis. *Eugen Sänger* accepted an offer to secretly help Egypt develop their own offensive missiles.

An intense-looking *Eugen Sänger* (left) about the time he was fired as director of West Germany's Research Institute for the Physics of Jet Propulsion (FPS) for his concurrent and on-going offensive rocket missile consulting services to Egyptian *President Nasser*.

Tank and a few others went to Egypt to give *Willy Messerschmitt* a hand on his Egyptian *Messerschmitt Me Ha 300* turbojet powered fighter after *Peron* fired them. But *Tank* was not ostracized like *Eugen Sänger* because the world community wasn't as concerned about military aircraft fighters as they were with offensive rockets and missiles.

Willy Messerschmitt **(1889-1970). He returned to aircraft design, development, and manufacturing in 1955 with aircraft exported to Egypt and Spain. Egypt purchased the turbojet-powered** ***Messerschmitt Me Ha 300*****. Like his colleague** ***Kurt Tank*****,** ***Willy Messerschmitt*** **didn't receive the world-wide public outcry working for** ***President Nasser*** **as what was heaped on** ***Eugen Sänger*****.**

Dr.-Ing. Irene Sänger-Bredt **continued her college-level teaching after** ***Eugen Sänger's*** **death. Critics claimed that she spent more of her valuable time writing popular histories of space travel and seeking to establish her husband's role in the pioneering space propulsion efforts than continuing on with her own well-respected rocket science work.**

Sänger also was an advocate of the *Einlauf* or a "pull in" type of diffusor, which held all the pressure rise inside the ramjet unit. His experiments were conducted using gasoline as fuel and he generally injected the fuel upstream, facing against the airflow, in order to obtain a reasonable efficiency of combustion. The combination of these factors resulted in a high drag ramjet unit because the combustion chamber normally had a length/diameter ratio of 5, while the entire ramjet's length was at least 10 times the diameter. *Sänger's* experiments were generally conducted with a high fuel/air ratio and consequently the specific fuel consumption was higher than the figures obtained by his competitor *Otto Pabst*. *Sänger's* experimental ramjet engines required approximately 8 grams of fuel per kilogram of thrust per second. He concentrated on the *Einlauf* diffuser because he intended to operate his propulsive ducts at *Mach* numbers close to unity and he was afraid that the *Fang* or "capture" type of diffuser advocated by *Dr. Klaus Oswatitch* of Göttingen University, would experience difficulty with local regions of supersonic velocity around the outer lip of the entry.

Oswatitsch was a research member of *Professor Prandtl's* Institute at Göttingen and had been working since 1943 on a ramjet-propelled artillery shell designed to be fired at supersonic speeds, of the order of *Mach 3*. He concentrated the majority of his efforts on the two problems of obtaining a supersonic diffuser with high pressure recovery and of developing an external shape for the com-

A ***Messerschmitt Me Ha 300*** **in Egypt. Notice in the far left lower corner is seen the nose of a Soviet-built** ***MiG*** **fighter aircraft.**

Dr.-Ing. Irene Sänger-Bredt **(center) with two long-time friends and supporters...*****Dr. Curt Debus*** **(left) and** ***Heinrik Seewald*** **(right). April 1975.**

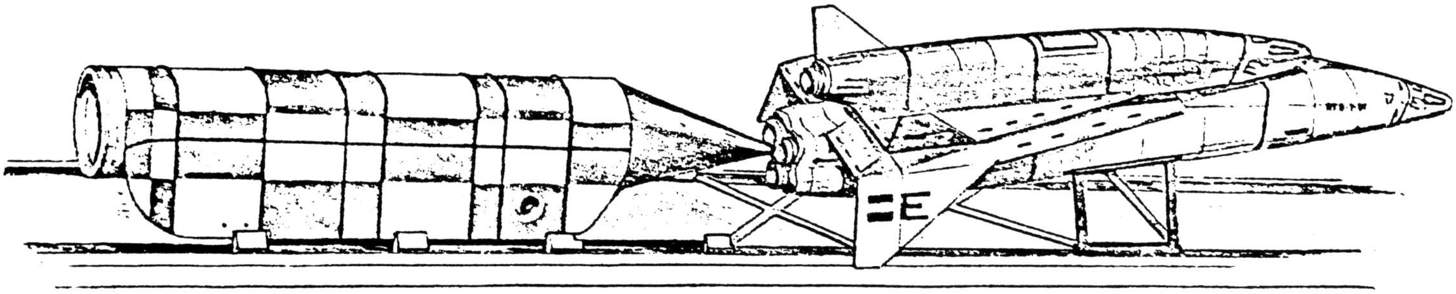

The *Space Transporter* (*Raumtransporter*) *Sänger II* proposed by *Junkers Flugzeug und Motorenwerke* (now a member of *MMB*) of West Germany in the early 1960s. Like *Eugen Sänger* and *Irene Sänger-Bredt* from the mid 1930s and early 1940s, a captive rocket booster accelerates the two-stage winged transporter to 560 miles/hour (900 kilometers/hour) along a 1.8 mile (3.0 kilometer) track and then brakes to a halt while the *"space transporter"* with its reusable orbital rocket-powered flying machine climbs away.

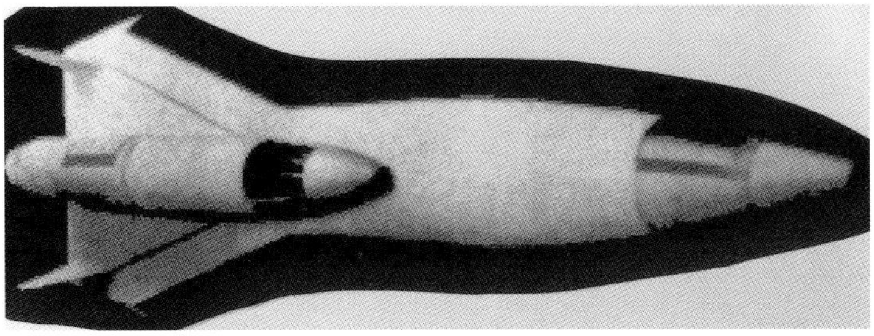

A scale model view of *Junkers Flugzeug und Moterenwerke's* proposed "mother carrier/booster" and its orbital, winged, and reusable *Sänger II* as seen from its front top. Launch weight would be 150- to 250-ton. Payload to low orbit would be between 2- and 3-ton.

bination missile and ram-jet which would have relatively low drag. It appears that *Sänger*'s *Lorin* duct ramjet models were not wind tunnel tested. Why this was not done is not known.. *Lorin* ducts were wind tunnel tested at *LFA* by *Dr. Knackstedt* under the direction of *Professor Busemann* . These tests were made on a 5 foot 7 inch [1.75 meters] model provided by *Helmuth Walter, Kiel (HWK)* and a much shorter model of only 1 foot 6 inch [500 mm] from *Otto Pabst* at *Focke-Wulf*. Modern day ramjet engines appear to have been more influenced by the theories and research work of *Oswatitsch* than *Sänger*.

The results obtained from the ramjet flight tests by *Sänger* at *DFS*-Ainring are rather difficult to analyze, because of the necessity of accurately measuring the thrust and drag and of separating the effect of the ram duct's supporting structure drag. The highest

An artist's concept drawing of the Aerospace Plane "Sänger II" leaving the runway. *Courtesy MBB-ERNO*.

The *Sänger II* separates from its "mother carrier/booster" approximately 30 km above the earth, off to fulfill its mission.

thrust coefficient (converted to the conventional basis of net available thrust) was 0.38 and a diffuser efficiency of 85 to 95 percent was apparently achieved. *Sänger's* forward flight speeds were never higher than 657 feet [200 meters] per second and the velocity ratio in the diffuser was approximately 6 to 1. It appears that *Sänger* made no attempt or didn't have to cool the ramjet fuel by water condensate film radiation, or any means other than external airflow and radiation. Internal gas temperatures, measured with a pyrometer during night flight tests, were of the order of 2000 degrees Centigrade.

Sänger provided approximately 80 fuel injection nozzles for his 4 foot 9 inch [1.5 meter] diameter propulsive duct and the pressure drop across these nozzles was calculated to 40% of the velocity head of the internal flow at that point. The diffuser had an inclined angle of 10 degrees divergence and the leading edge was slightly rounded to provide for smooth entry flow conditions. The

The larger vehicle, "mother carrier/booster" would return immediately to land on a conventional runway, while the smaller orbital vehicle or *"Sänger II"* would continue to climb to orbit. After the mission the orbital *"Sänger II"* - constructed largely of nickel steel with ablative coatings - would make a glide reentry into the Earth's atmosphere.

Illustration of *Junkers' Flugzeug und Motorenwerke* proposed *Sänger II* project featuring how the orbital vehicle would land after returning from space. *Junkers'* engineers proposed that the orbital machine land on its tricycle ski-type undercarriage instead of wheels. *Eugen Sänger* was a consultant for *Junkers'* on their *Sänger II.*

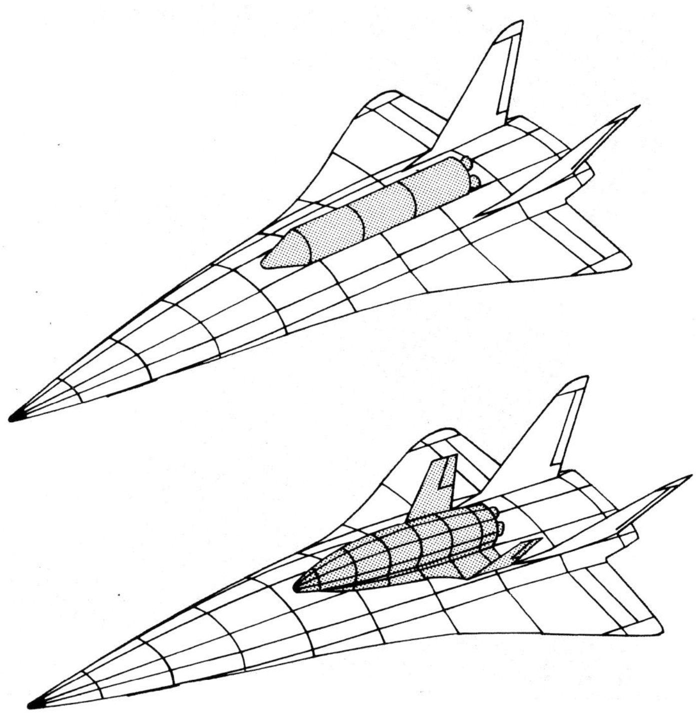

Two pen and ink versions of the "*Sänger Projekt*" of *Messerschmitt, Bölkow, Blohm* (*MBB*) of West Germany, which included a "mother carrier/booster" and a reusable orbital in the form of a transporter or a passenger carrying flying machine. *MBB* named their concept in honor due to similar ideas by *Irene Bredt-Sänger* and *Eugen Sänger* from the 1930s and 1940s.

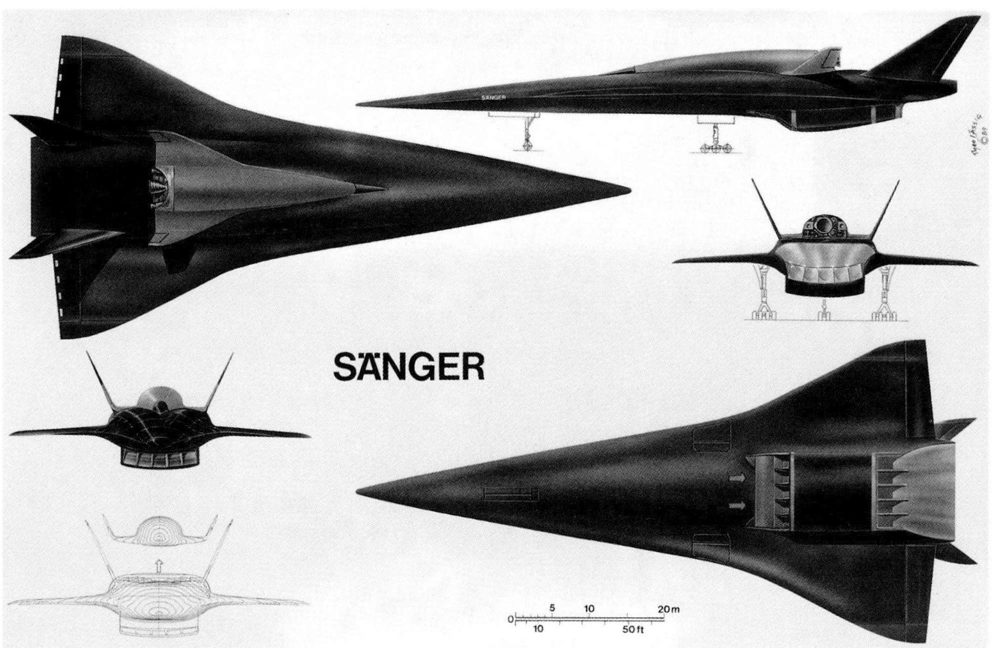

Multiple views of *MBB's* "*Sänger Projekt,*" which was aptly named, given its obvious use of *Sänger's* previous research.

***MBB's "Sänger Projekt"* was designed to be capable of taking off and landing at Europe's airports, negating the need for any sort of launch pad or rail as in earlier designs by *Sänger*.**

exit area was established so as to produce an entry flow velocity of approximately 492 to 590 feet [150 to180 meters] per second with complete combustion, and the major portion of the flight work was accomplished at a flight speed equal to this duct entry velocity.

When *Sänger* began conducting his ramjet combustion trials at the Experimental Aircraft Point in 1941, his first ramjet to be tested was nicknamed "the flying stove pipe." For his combustion trials tests, the 5 foot long prototype ramjet was mounted on a 2 1/2-ton *Opel "Blitz"* truck and used as a moving test bed. *Sänger* found that his ramjet functioned according to his theories, however, the speed of the truck was too slow to compress the inflowing air sufficiently to allow a propulsive thrust to be measured. *Sänger* realized that the truck test bed would have to be replaced with a test bed which flew and he somehow was able to obtain a *Dornier Do 17 Z-2* (with the designation *BC+NL)* – a twin-engined bomber – along with a pilot by the name of *Paul Spremberg* from the firm *Rheinmetall-Borsig*.

Sänger continued his ramjet flight testing in the more friendly atmosphere of *Professor Georgii's DFS*-Ainring in a department known as the Institute for Special Propulsive Units. It is not clear if the *Do 17 Z-2* which *Sänger* had started using at Trauen/Fassberg in 1941-42, followed him to *DFS*- Ainring. Nevertheless, *DFS* had a *DornierDo 217K V3* at Ainring and it may have shared the ramjet flight testing with other *DFS* duties such as carrying the *Me 328* pulse-jet powered miniature fighter and *DFS'* own *DFS 228* bi-fuel rocket-powered ultra high altitude reconnaissance prototype. The

The *Sänger* aerospace plane by *MBB* would be able to take the transport to all Low Earth Orbits (LOEs), and then launch the transport for further travels. The photo at right depicts the *Sänger* and transport approximately 35 km above the earth's surface.

The USAF's *Glenn L. Martin Aircraft Company's X-24B.* It was constructed to investigate reusable lifting body designs from Edwards Air Force Base in the mid 1970s to determine the best way of bringing astronauts home from space, instead of in capsules that simply fell out of the air. *Martin's X-24B* reusable lifting body had vestigial wings and a flat underside similar to *Eugen Sänger's* orbital rocket bomber project of the early 1940s, however, *Glenn L. Martin's X-24B* was a blended body, that is, it was hard to say where wing stopped and fuselage began. The challenge was to construct a flying machine which might reach 5,000 miles/hour yet flyable at 200 miles/hour in order to make it slow enough to land.

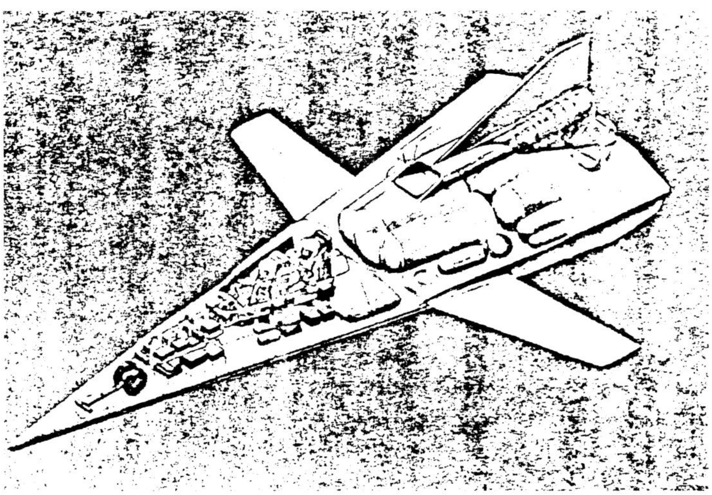

A poor quality pen and ink drawing of *Lockheed's* 2-man lifting body concept with variable-wing geometry "swing wing" to assist landing.

method used in *Sänger's* ramjet flight testing was to mount the ramjet on the back of the *Dornier Do 217K V3.* The *Do 217K* would take off and gain altitude under its own piston-engined power until it reached approximately 22,000 feet. The ramjet would be tested at this height, and lower altitudes as well, and *Sänger* reported constant operations of 10 to 30 minutes in level flight. For the most part, the tests were done safely with only a few minor mishaps. Ramjet operational tests were conducted, also, in a power dive. Pilot *Spremberg* would take the *Dornier Do 217K V3* up to 13,000 feet altitude before going into a shallow dive. With the ramjet turned on the *Dornier Do 217K* reportedly reached up to 420 miles/hour in the diving tests. Individuals who witnessed these diving tests with the ramjet fixed atop the *Dornier Do 217*, claimed to have never had seen anything like it before. It must have been a sight, indeed, with this roaring, flaming ramjet on top of an aircraft in a screaming power dive of speeds reaching 420 miles/hour.

Sänger was one of the first to suggest the use of fuel blocks prepared from coal dust and plastics for use in propulsive ducts to be installed in several of *Alexander Lippisch*-designed proposed 1944-1945 fighter aircraft. These coal dust blocks were to be bonded to the inner lining of the walls of the duct. This method, said *Sänger*, solved the problem of feeding fuel into the combustion space of the duct, which in many cases, was not a round duct anymore but more like a rectangular duct. In the early to mid 1940s, *Dr. Johannes Winkler*, the former first President of the *VfR*, was researching coal-dust based fuel blocks at *LFA*-Braunschweig. These also were being proposed for *Lippisch's* ramjet-powered fighters. It appears that *Winkler* was making more progress on fuel blocks that *Sänger* and that *Winkler* was the leading expert in this field at war's end. *Otto Pabst*, of *Focke-Wulf* succeeded in reducing the length of such ducts for use on his proposed *Focke-Wulf Fw Ta 283* interceptor/fighter and his vertical take-off *Focke-Wulf Fw Triebfugel. Pabst* became interested in ramjet engines after witnessing *Eugen Sänger's* experiments atop a *Dornier Do 217. Pabst* believed that *Sänger's* ramjet concept was much too long to be practical and that he could design a shorter ramjet engine. However, the *Lorin*-type propulsive duct engine had one great disadvantage – it did not produce enough thrust at low velocities. Also, there was no possibility of accelerating a missile from zero velocity, or from a low speed, to transonic or supersonic velocities by using this type of propulsion. The liquid rocket engine was free from this disadvantage and, at the expense of much higher fuel consumption, very high accelerations were possible at any velocity.

The brand new *Glenn L. Martin Aircraft's X-24B* shortly after arriving at Edwards Air Force Base. Its fuselage shape was similar to the *Sänger* sub-orbital bi-liquid fueled rocket bomber.

***Eugen Sänger* and friend *Ferdinand Brandner* (right). *Brandner* was the former chief of engine gas piston development at *Junkers Jumo*-Dessau. He was arrested by the Soviets and charged as a war criminal. Imprisoned for two years in the USSR, he was released to help work on the Soviet version of the *Jumo 022K* turboprop at Kuibyschev. Photo taken in France in the mid to late 1950s. *Brandner* was fond of reminding his friend that a prophet is seldom recognized in his own country and time.**

The Germans realized turbojet engines could be used quite effectively on subsonic aircraft. However, given the state-of-the-art in turbojet engine development in the early 1940s, the *RLM* wasn't exactly sure how efficient these engines – which mechanically compressed the air required for their combustion/thrust – might be beyond the speed of sound. The *RLM* also argued that at very high forward speeds of more than 1,100 miles/hour, the necessity of mechanically compressing the incoming air becomes moot, since the necessary pressure for combustion can be obtained from the aircraft's forward velocity itself. This was the nature of the discussions between *Sänger* and officials from the *RLM*. *Sänger* advocated that an air-breathing motor could dispense with the necessity of having a turbine with its mechanical compressor. Doing away with the heavy weight and mechanical complexity of a spinning compressor and turbine considerably simplified the construction of an aero motor. *Sänger* said this was the entire nature of the so-called ducted air breathing reaction motor, or *Lorin* duct. He believed that the *Lorin* duct could even be placed in aircraft frames for subsonic performance. What was not fully understood was how much loss in performance would result if the *Lorin* duct propelled airplane experienced steadily slower and slower velocity at its air intake? Critics of the *Lorin* duct for aircraft usage at the time believed that optimum operating velocities were in the nature of 1,850 to 3,100 miles per hour. Therefore these units were not considered worth the expenditure of time and money spent to more fully understand the true nature of this propulsion concept. Based on the initial discussions between *Sänger* and a group of powerful leaders within the *RLM* s such as *Oberst Siegfried Knemeyer* and *Helmut Schelp*, it was agreed that *Sänger* would set about investigating the possibilities of a *Lorin* duct being installed in *Luftwaffe* fighter aircraft. *Siegfried Knemeyer* was the leader of this so-called "club" and from 1943/1944, they decided what research would or would not be pursued by the *RLM*. It was *Knemeyer* and his club, for example, who authorized the construction of the speed-of-sound record breaker *DFS 346* to be built in early 1945... just to have 'bragging rights' with Allied victors after the war.

Postwar, *Eugen Sänger* theorized a space ship circumnavigating the cosmos in a photon rocket. It could also be called 'propulsion by light rays'. *Sänger's* hypothesis was that a reaction type propulsion could be generated by energy expelled at the speed of light...186,000 miles/second. Sun light falling on each square yard of Earth, for example, produces a measurable pressure on the surface of the planet, even though it may only be about 1 milligram.

***Eugen Sänger*, 1959, at 54 years.**

***Hartmuth Sänger* at *Trauen/Fassberg*, July 2001.**

"Space travel must come; it follows the development of land, sea, and air travel with the inevitability of a natural law, just as the atomic bomb followed gunpowder and dynamite. Do not believe that we are so arrogant as to claim to decide whether it ought to be carried out or not; our intellects are only tools for testing means – the orders were given to us by nature through our own inner impulses. Do not believe, either, that consideration of profit or loss will have the last word in space travel; the decision will be made by the elemental joy in great deeds and by that alone, for nothing else but the revelation of the secrets of the unknown can give meaning to life. It was this which led (Admiral) *Byrd* and *Amundsen* to the Poles, which sent (Colonel) *Lindbergh* across the Atlantic and led others to climb the highest mountains in the world. It is this zeal which inspires those who are working towards space travel in the face of difficulties and disappointments and which – together with our limited insight into the workings of nature – will eventually bring us success and the power to direct the great exploit with wisdom for the common good. Flight in cosmic space is now being prepared as humankind's next step in the fulfillment of our *Faust*-like will to strike out beyond the frontiers of knowledge."

However, if we take a large area – the Atlantic Ocean, for example – sunlight puts a pressure on this body of water equal to about 107,000-ton because the area subject to radiation is so large. Were this radiation to be concentrated in the form of an engine about the size of one of *Sänger's* ramjets it should develop enough power to propel an airplane. The problem is one of concentrating this thrust. *Sänger* believed that the photon rocket engine would be the ultimate propulsion unit – the form of propulsion needed to power humankind out into interstellar space and back. He was fond of saying we'd have to be careful how fast and how far we traveled. For example, *Sänger* considered a space ship circumnavigating the cosmos. During his life, he and others assumed the distance across the breadth of the universe to be roughly 10 billion light-years. *Sänger* said that if the ship could achieve 99.999 percent of the velocity of light, then the crew would imagine that their journey had lasted only 33 years – yet 10 billion years would have elapsed before they returned to Earth...if it still existed, and if they could still find it. *Arthur C. Clark* wrote (*The Promise of Space*) that since this feat of space travel would require the complete conversion into energy of a mass approaching that of the Moon, *Sänger* decided that this far surpassed the limits of the technically feasible. But *Clark* felt that *Sänger* might be right. "Because we now know," he said, "that there are energy sources in the universe which appear to turning Moon-sized masses into radiation every 10 seconds, it might be best to reserve judgment even on this point."

When *Sänger* accepted West Germany's offer to head up the Institute of Jet Propulsion Physics- Stuttgart University in 1954, he said:

***Hartmuth Sänger* at *Trauen/Fassberg*, July 2001.**

The Rocket Bomber in Flight

Planet Earth as seen by a pilot of the *Sänger* winged orbital rocket bomber as he/she levels off at about 80 miles altitude. Digital image by *Mario Merino*.

The port side nose, to give underside of a *Sänger* winged orbital rocket bomber appearing across the wind screen of an observing flying machine. Digital image by *Mario Merino*.

The simplified underwing *Balkenhreuz* (*B4* style) signifying 1944-1945 *Nazi* Germany's *Luftwaffe,* appears in view as the *Sänger* winged orbital rocket bomber continues passing by the observer. Visible, too, is the flying machine's closed rectangular-shaped bomb bay doors. The two smaller rectangular-shaped boxes to its port and starboard are the winged orbital bomber's main landing gear doors. Digital image by *Mario Merino*.

Coming into view is the *Sänger* winged orbital bomber's starboard side rectangular-shaped vertical stabilizer with a hinged rudder. Digital image by *Mario Merino*.

The *Sänger* winged orbital bomber's starboard side rectangular-shaped vertical stabilizer with a hinged rudder. The entire outboard side of the stabilizer is painted red with a black *Halkenkreuz* (*swastika*) filling the space between the stabilizer's leading and trailing edges. Digital image by *Mario Merino*.

The thrust end of the *Sänger* winged orbital rocket bomber and featuring its main 100-ton thrust bi-fuel liquid rocket engine. The outboard nozzles were for limited space maneuvering. Digital image by *Mario Merino*.

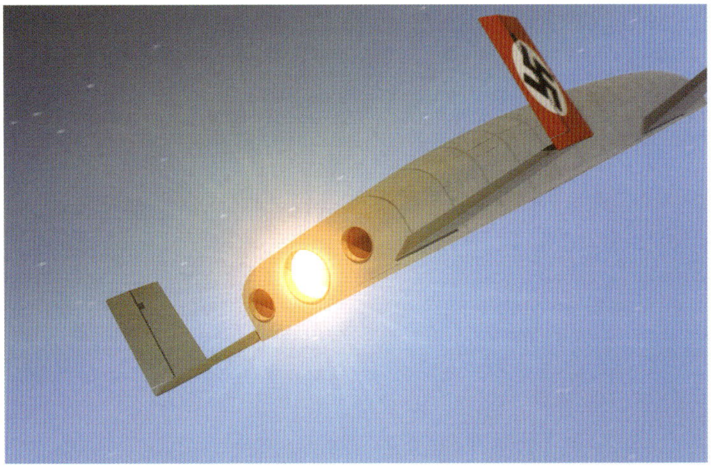

The *Sänger* winged orbital bomber from its rear starboard side and featuring its dual vertical stabilizers with hinged rudders. Elevator controls were installed in the horizontal stabilizer. These controls could only be effective while in the atmosphere...during climb and later when landing. Digital image by *Mario Merino*.

The *Sänger* winged orbital bomber, with its 100-ton rocket engine burning, appears approaching the end of its powered flight. Digital image by *Mario Merino*.

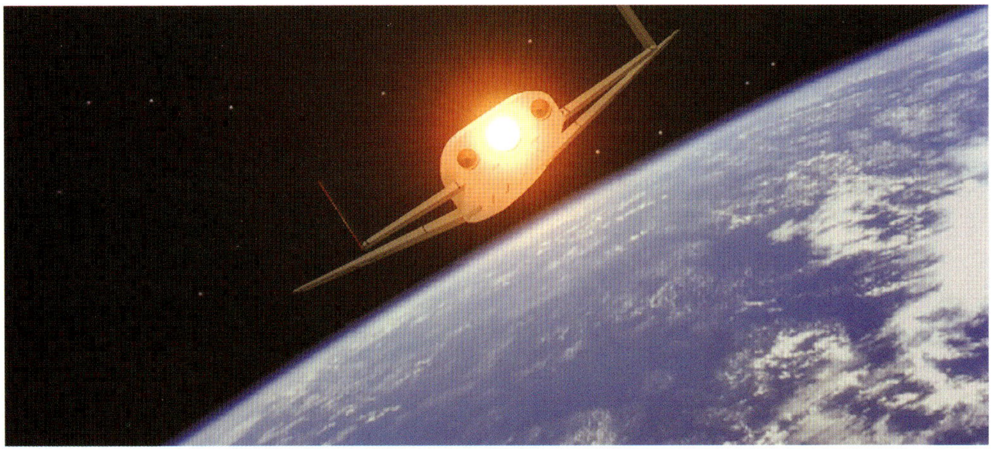

The *Sänger* winged orbital bomber, its 100-ton rocket engine still burning, could not have gone into an elliptical or circular orbit. Upon burning all its 90-ton of fuel, the *Sänger* would continue around the earth sub-orbital in a series of undulating skips. Digital image by *Mario Merino*.

As early as the Summer of 1938, *Sänger* outlined the development and testing program for his winged rocket plane in the journal *Flug*. *Sänger's* winged rocket plane is seen approaching almost head on. Digital image by *Mario Merino*.

Only a small number of individuals were pursuing winged rocketry in Germany in the 1930s. In April 1931, *Reinhold Tiling* (1875-1934) of Osnabrück demonstrated his latest winged rocket. It consisted of a streamlined aluminum body housing the rocket. It featured two tail fins while its wings snapped out like the blade of switch-blade knife. Digital image by *Mario Merino*.

The *Sänger* winged orbital rocket plane. He called it his plano-convex supersonic glider according to his German patents (April 22nd 1939), with a half ogival-shaped fuselage, nose, and wedge-shaped wings. Digital image by *Mario Merino*.

Because of its dome-shaped body profile and flat bottom, *Sänger's* assistants nicknamed the supersonic winged rocket glider the "*flat iron*." Digital image by *Mario Merino*.

Sänger's winged reusable sub-orbital rocket bomber has been known by a variety of names including: *Glide Bomber*, *Boost Glide Bomber*, *Skip Glide Bomber*, and the *Antipodal Bomber*. Digital image by *Mario Merino*.

The name *Antipodal Bomber* is derived from the world's antipodes, which are simply points on the opposite sides of the earth, such as Berlin and Australia/New Zealand. Digital image by *Mario Merino*.

Sänger **also called his winged orbital bomber "*Robo*" meaning *Raketenbomber*. Digital image by *Mario Merino*.**

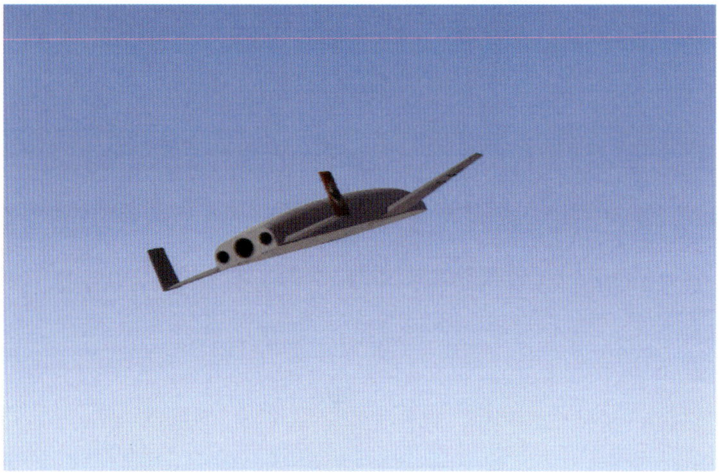

Sänger's "Robo" **was an Earth-orbiting single-stage rocket plane with a launch weight of 100-ton (90- ton of propellant and payload). Digital image by *Mario Merino*.**

Maximum loaded performance for the winged orbital bomber to its antipodal point on the Earth was calculated up to 8-ton of bombs. Digital image by *Mario Merino*.

Sänger **wrote that he wanted to chose for the body the shape of a projectile, pointed in front and blunt at the back-end to give room for the exhaust velocity. Digital image by *Mario Merino*.**

The semi-ballistic shape of the *Sänger* winged orbital bomber is evident when it is viewed from this perspective. Digital image by *Mario Merino*.

Eugan Sänger **wrote that the wing profile for his winged orbital bomber had to be as thin as possible, with sharp leading edges. Thus, the wing span can then be kept low because of the negligible resistance of the wing-edges. Digital image by *Mario Merino*.**

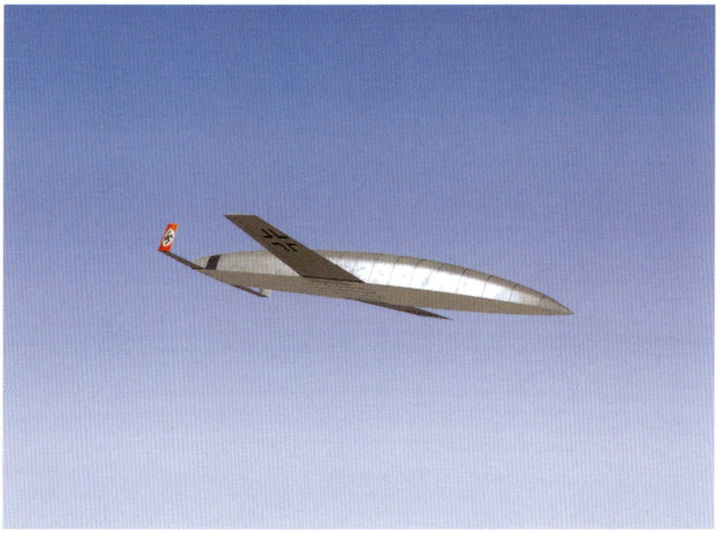

Regarding the wings and the lift provided, *Sänger* said that the forces of inertia of the orbit would contribute to the winged bomber's lifting effect. Digital image by *Mario Merino*.

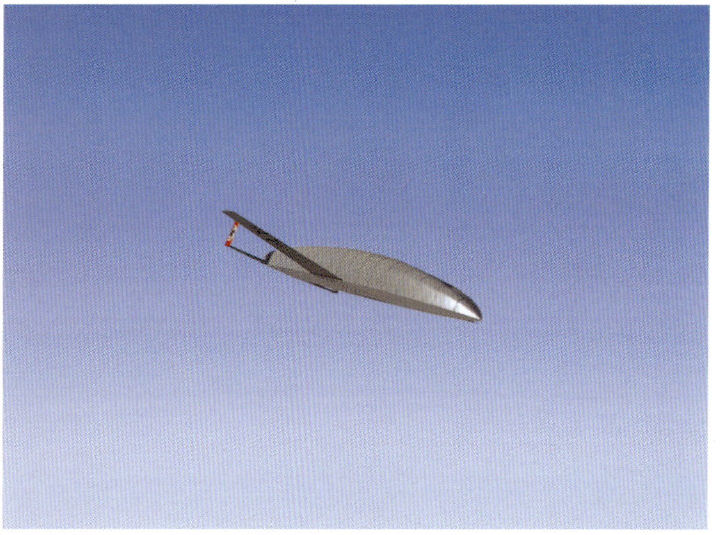

Eugan Sänger said that once a winged orbital rocket bomber such as his reached 18,020 miles/hour [29,000 kilometers/hour] the centrifugal force of the curved trajectory would be equal to the weight of the plane so the wings wouldn't need to provide lift any more. Digital image by *Mario Merino*.

Sänger calculated the mean wing lift-drag ratio on his winged orbital bomber was 5, and an effective exhaust velocity of 12,139 feet/second [3,700 meters/second]. Digital image by *Mario Merino*.

We are back to the nose, completing a full visual circle around the winged *Sänger* sub-orbital bomber. Digital image by *Mario Merino*.

Sänger wrote a detailed engineering report on his proposed winged orbital bomber. He presented it to colleagues, academics, and *Luftwaffe* senior officers at the *RLM* on March 17th 1941. *Sänger's* proposal received outright rejection. Digital image by *Mario Merino*.

A see-through image of the *Sänger* winged orbital bomber and featuring the pressurized cockpit cabin, liquid oxygen tanks, fuel tanks, and the winged rocket's combustion chamber and exhaust nozzle. Digital image by *Mario Merino*.

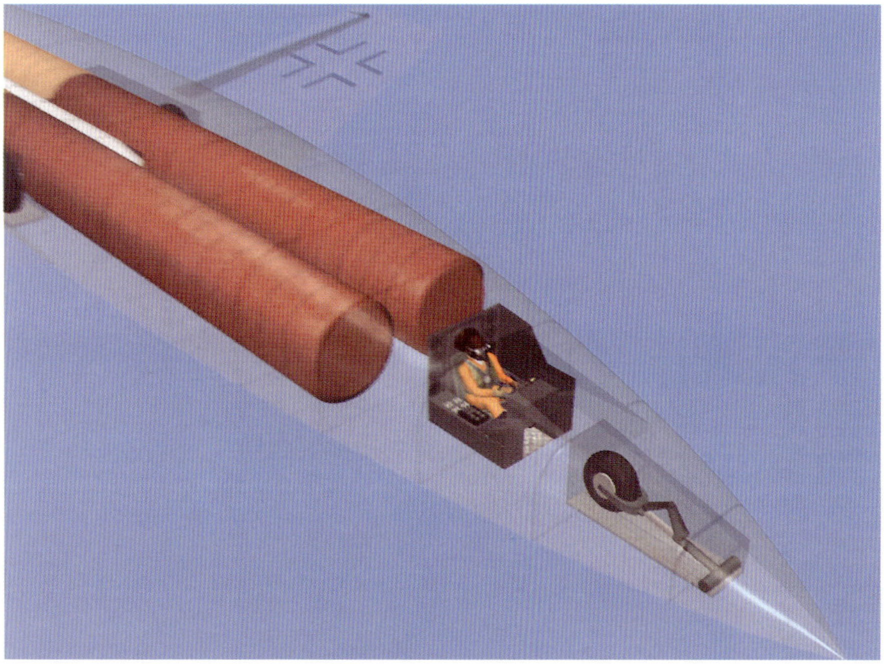

Left: Featured is the *Sänger* winged bombers' retracted nose wheel and the pressurized single seat cockpit cabin. Directly behind the pilot's cockpit are the two liquid oxygen tanks. Digital image by *Mario Merino*.

Below: The pilot's pressurized cockpit cabin as seen from its starboard side. Digital image by *Mario Merino*.

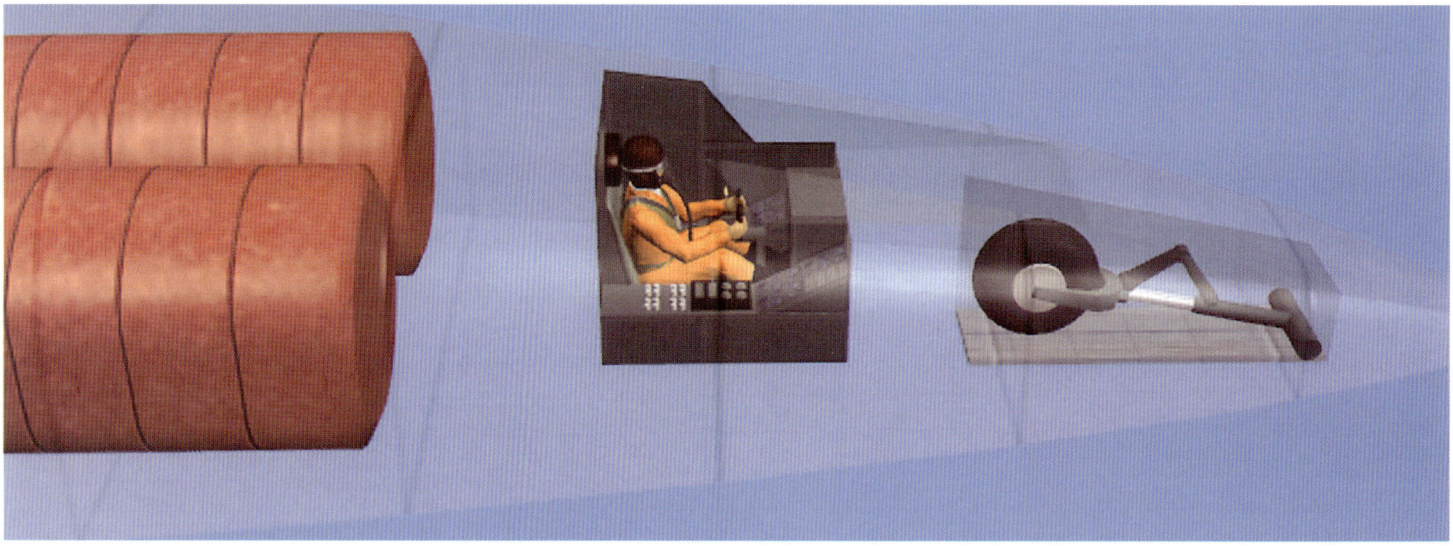

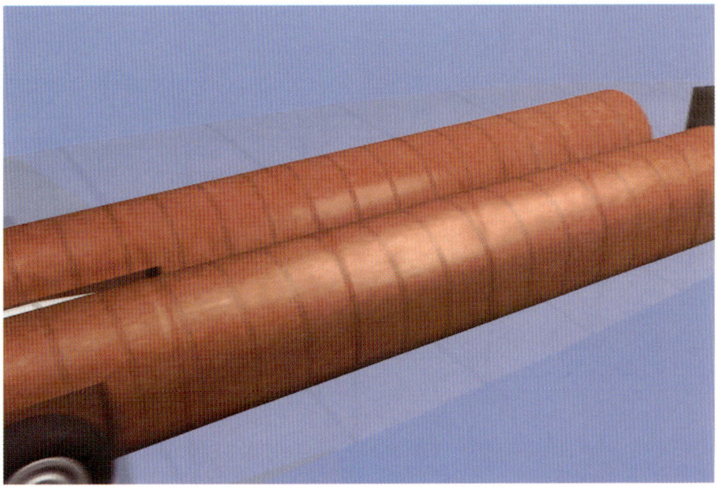

Starboard side view of the winged rocket bomber's twin liquid oxygen tanks. Digital image by *Mario Merino*.

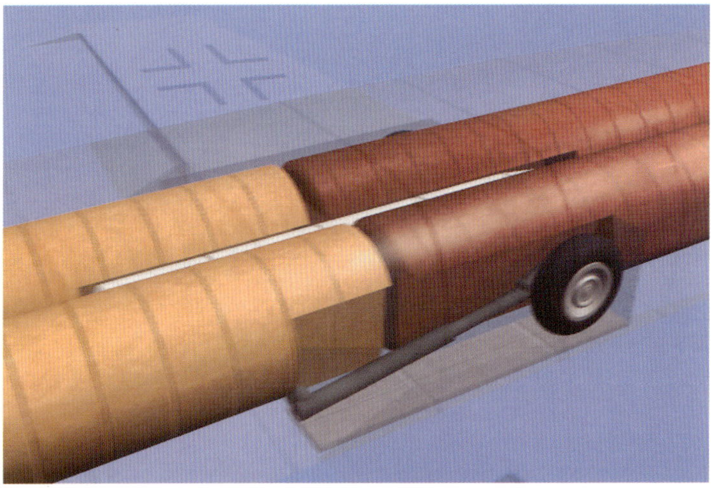

Position of the starboard side main landing gear. Notice that both the starboard side lox tank as well as the fuel tank had to give up space to allow the landing gear to articulate. Digital image by *Mario Merino*.

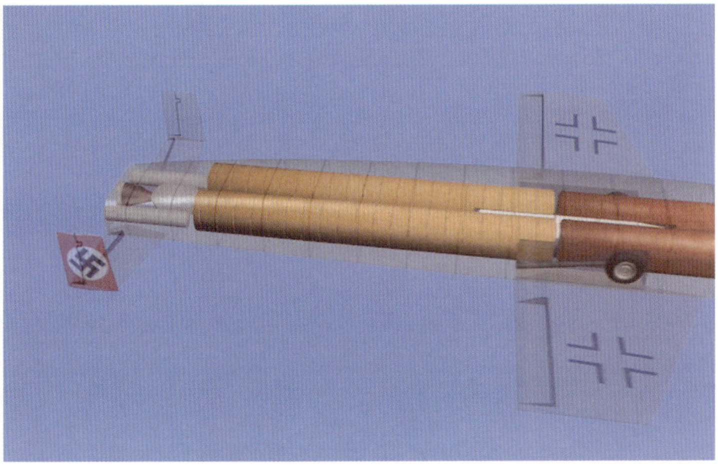

Starboard side internal view of the winged rocket bomber's fuel tanks...directly aft the liquid oxygen tanks. Digital image by *Mario Merino*.

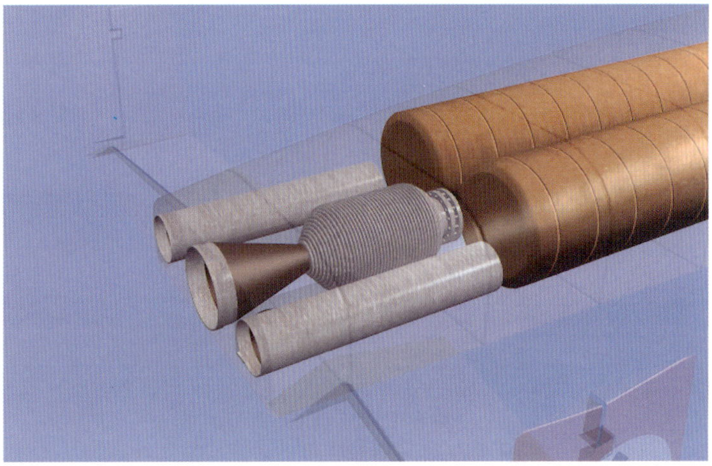

An internal view of a winged bomber's 100-ton thrust combustion chamber and exhaust nozzle. Liquid oxygen would have been injected into the combustion chamber at a pressure of 100 atmospheres. Digital image by *Mario Merino*.

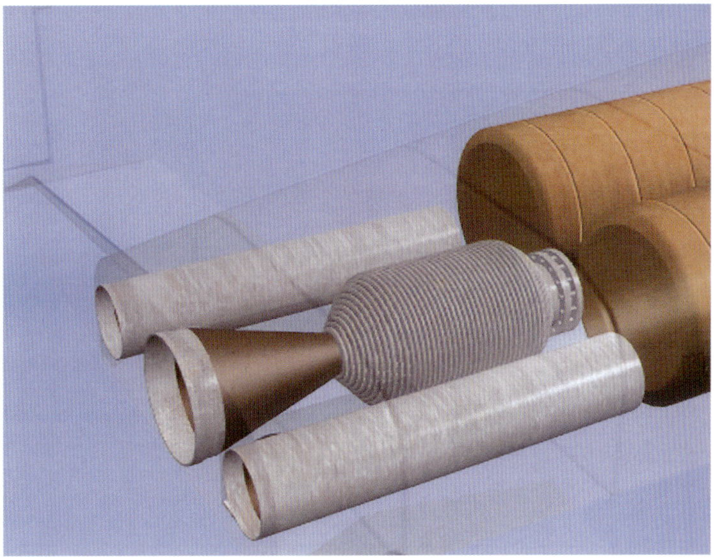

***Eugen Sänger* had been working on a full scale model of the winged bomber's 100-ton thrust bi-fuel liquid rocket engine, however, his rocket engine research work at Trauen/Fassberg was terminated by *Hermann Göring* in Autumn 1942. Later *Göring* learned that he had been wrong, and that the German Army's *A-4 (V2)* was not the threat to his *Luftwaffe* as he had initially feared. Digital image by *Mario Merino*.**

The complete launch program...the *Sänger* winged sub-orbital rocket bomber atop its 1.86 mile long monorail horizontal launch, and the propulsive shed containing a number of of 50,000 pound thrust each *A-4 (V2)* rocket engines. Digital image by *Mario Merino*.

The propulsive sled containing thirty *A-4 (V2)* 50,000 thrust rocket nozzles which current research suggests would be required. *Sänger* did not estimate the exact number of 50,000 thrust *A-4* engines required to push his winged orbital rocket bomber up to a lift off speed nearly 1,200 miles/hour at the end of 1.86 mile-long launch. Digital image by *Mario Merino*.

Ignition sequence bring life to the 30 *A-4 (V2)* 50,000 pound thrust rocket engines. Digital image by *Mario Merino*.

The 30 *A-4 (V2)* 50,000 pound thrust rocket engines are now at full thrust and the winged bomber begins to inch again. Digital image by *Mario Merino*.

More movement as the wall of energy intensifies. It would take the propulsive sled 11 seconds to reach approximately 1,075 miles/hour takeoff speed by the end of 1.86 miles. Digital image by *Mario Merino*. *Sänger* estimated the combined weight of his winged bomber and its propulsive sled to be approximately 150-ton. Unlikely. Digital image by *Mario Merino*.

Imagine the roar of 30 *A-4* (*V2*) rocket engines at full thrust speeding down the 1.86 mile long monorail. Digital image by *Mario Merino*.

A view of the winged bomber and its propulsive sled as seen from in front and above. Digital image by *Mario Merino*.

Acceleration is mounting steadily as the winged bomber and its propulsive sled head for the end of the 1.86 mile long monorail launch track. Digital image by *Mario Merino*.

A view of the winged bomber and its propulsive sled down range from its starting point. Digital image by *Mario Merino*.

The end of the 1.86 mile long monorail can be seen, and the package has reached its top speed of nearly 1,100 miles/hour. Digital image by *Mario Merino*.

The *Sänger* winged rocket bomber has lifted off its cradle and is moving away. Digital image by *Mario Merino*.

Seen from its nose port side, the winged bomber's forward motion is benefitting from kinetic energy pushing it off the end of the monorail plus its own 100-ton thrust bi-fuel liquid rocket engine. It leaves its launch ramp at a 30 degree angle to the ground. Digital image by *Mario Merino*.

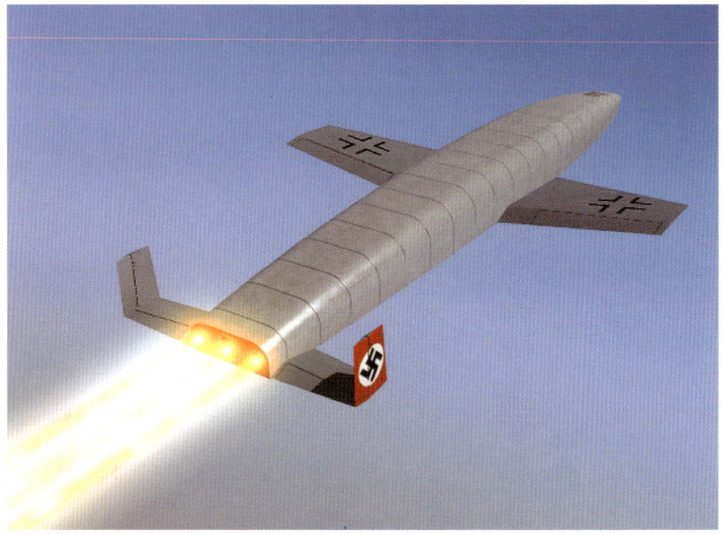

The ballistic shape of a *Sänger* winged bomber with its 100-ton thrust engine creating a whitish yellow exhaust trail immediately behind its thrust nozzle. Digital image by *Mario Merino*.

A *Sänger* winged bomber's two outboard maneuvering solid fuel rocket engines augment the 100-ton center thrust nozzle. The bomber has to be alined perfectly while still at low altitude so that its flight path will match the course needed for its orbital flight.Digital image by *Mario Merino*.

Still under its own power, the winged *Sänger* rocket bomber is coming toward us just before the sun is setting. Digital image by *Mario Merino*.

The *Sänger* winged rocket bomber is leveling off at its first peak of about 162 miles altitude. Digital image by *Mario Merino*.

The winged rocket bomber reflecting the setting sun off its 92 foot long starboard side fuselage. Digital image by *Mario Merino*.

The *Sänger* winged rocket bomber has overtaken our observation place and speeds by. Digital image by *Mario Merino*.

Upon reaching 162 miles [261 kilometers] altitude, the *Sänger's* altitude would immediately deteriorate...eventually gliding down to its first low at 25 miles [40 kilometers]. The dense atmosphere begins at about 25 miles altitude. Digital image by *Mario Merino*.

A *Sänger* winged orbital rocket bomber's altitude is deteriorating rapidly. However, it will continue to glide down to Earth for a few more moments before reaching about 25 miles [40 kilometers] altitude. Digital image by *Mario Merino*.

The *Sänger* winged rocket bomber is still falling but not quite yet down to its first low peak of 25 miles [40 kilometers] altitude. Digital image by *Mario Merino*.

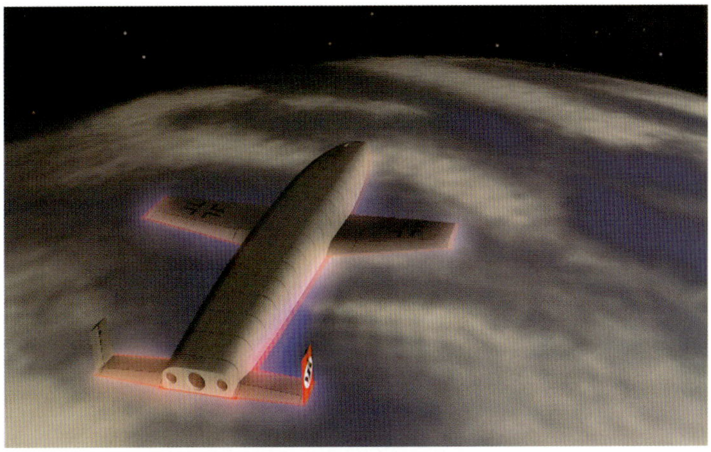

The denser atmosphere at about 25 miles altitude [40 kilometers] creates friction, and the leading edges of the *Sänger's* tail assembly are glowing red as it "bottoms out" and again experiences the sensation of being propelled higher up through the dense atmosphere. The winged rocket bomber has traveled 2,796 miles [4,500 kilometers] from its launch site in *Nazi* Germany. Digital image by *Mario Merino*.

The unpowered *Sänger* begins to skip off the denser atmosphere, an action which propels it up again toward its 2nd peak of 78 miles altitude. Its wings leading edge surfaces are glowing red due to atmospheric friction. Digital image by *Mario Merino*.

Gliding up through the dense atmosphere friction creates additional heating, and much of the winged bomber is now glowing more...than before. Digital image by *Mario Merino*.

The winged bomber is approaching its 2nd peak of 78 miles [125 kilometers] altitude and begins leveling off. The fuselage leading edges are still glowing red hot as its pilot now makes arrangements to begin his/her bomb release. Digital image by *Mario Merino*.

The *Sänger* winged bomber's leading edges are cooling down rapidly at its 78 mile altitude. The winged bomber flight attitude is level. Digital image by *Mario Merino*.

Everything is quiet around the winged *Sänger* rocket bomber. The pilot is checking his/her bomb trajectory calculations. Digital image by *Mario Merino*.

The *Sänger* winged bomber has started crossing the Pacific Ocean...78 miles beneath...a cloud obscures this milestone along its round-the-world flight. Digital image by *Mario Merino*.

Gliding forward the obscuring cloud cover disappears just as the *Sänger* piloted, winged, sub-orbital rocket bomber approaches the west coast of the USA in the vicinity of San Francisco. Digital image my *Mario Merino*.

Gliding, proceeding forward, the piloted, winged, sub-orbital rocket bomber approaches its bomb release point...about 78 miles above Indianapolis, Indiana. Digital image by *Mario Merino*.

A single bomb bay door begins opening at about 78 miles above Indianapolis, Indiana as the winged bomber silently glides by high overhead unseen. Digital image by *Mario Merino*.

A scattered cloud cover remains as the winged *Amerika Bomber's* bomb bay doors continue opening revealing its radiation-spread bomb. Digital image by *Mario Merino*.

Bomb bay doors are full open now on the *Sänger* winged sub-orbital rocket *Amerika Bomber*. Digital image by *Mario Merino*.

Decision time. The pilot releases the 5,000 pound high explosive bomb which has been wrapped in blankets of radioactive silica (sand). He probably does not see it as it falls away to earth. Digital image by *Mario Merino*.

The long cylindrical object has cleared the winged orbital *Amerkia Bomber* as the flying machine continues steadily onward. Digital image by *Mario Merino*.

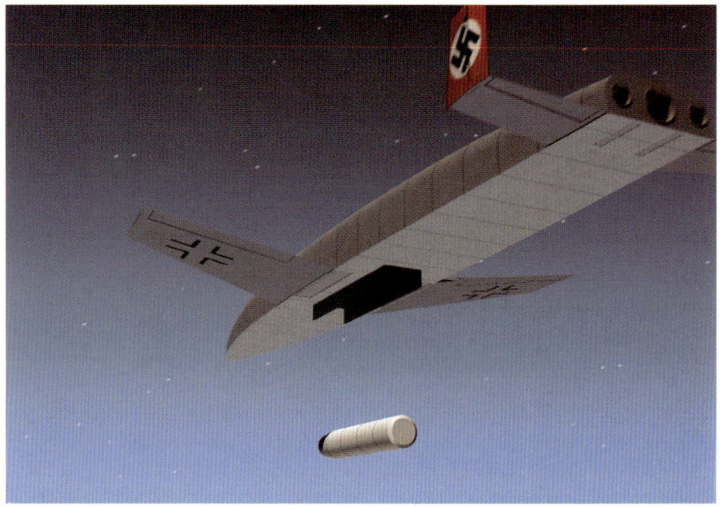

The width and length between the winged bomber and her bomb steadily increases, and the radiation-spreading bomb is being slowed down by atmospheric friction on its way to New York City. Digital image by *Mario Merino*.

The winged orbital bomber pilot is making plans for a touch down in *Nazi* Germany as the radiation-spreading bomb falls lower and further behind the flying machine. Digital image by *Mario Merino*.

The winged *Sänger* orbital rocket bomber is long gone above New York City heading home across the Atlantic Ocean as the *Nazi* radiation-spreading bomb passes through the scattered cloud cover. Digital image by *Mario Merino*.

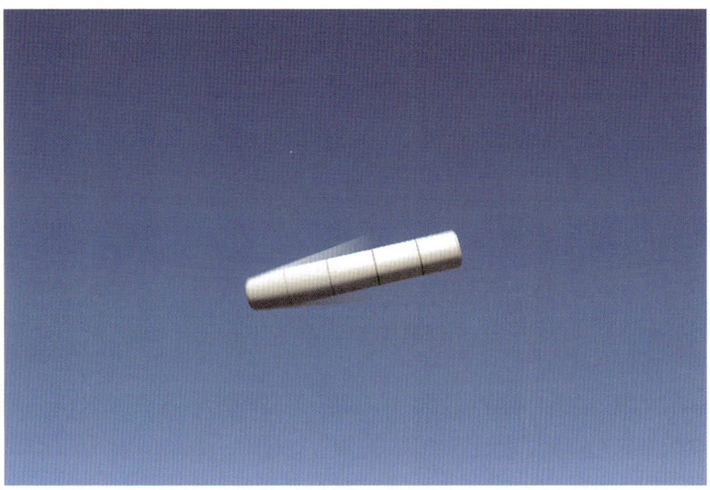

Flying through the scattered cloud cover...the *Nazi* radiation-spreading bomb has slowed its forward speed and is within a few miles above New York City. Digital image by *Mario Merino*.

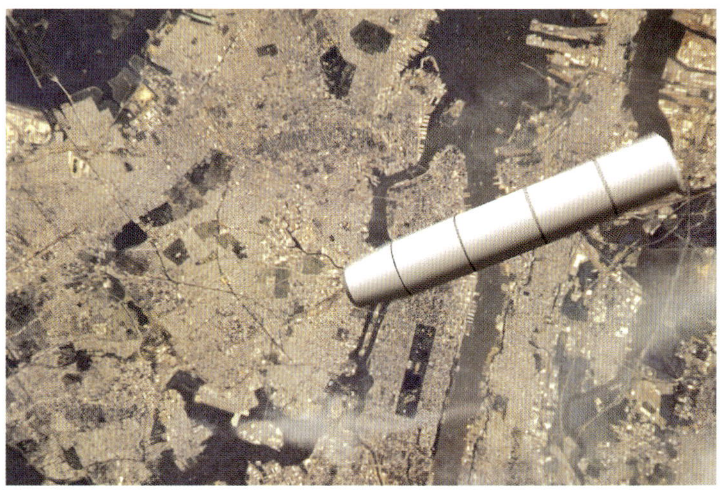

The white cylindrical-shaped *Nazi* radiation-spreading bomb is now tumbling down toward Manhattan Island...ground zero. Digital image by *Mario Merino*.

Manhattan Island's East River and Central Park are coming closer as the *Nazi* radiation-spreading bomb is silently falling out of the sky. Digital image by *Mario Merino*.

Between 2,000 and 3,000 feet the *Nazi* radiation-spreading bomb self detonates...across the East River from Manhattan Island over the borough of Queens. Digital image by *Mario Merino*.

Right: The spreading ring of radiation-spreading fire has now engulfed all of metropolitan New York City and Newark, New Jersey. Digital image by *Mario Merino*.

A small flash of light is now growing brighter, spreading increasingly outward in a ring of fire. Digital image by *Mario Merino*.

A well-defined ring of fire is growing, spreading outward covering almost all of New York City's five boroughs, as well as across the Hudson River into northeastern New Jersey. Digital image by *Mario Merino*.

Outward, the spreading ring of radiation-spreading fires continues its flight as if seeking to run away from itself. Digital image by *Mario Merino*.

The *Nazi* ring of fire has extended outward about as far as it will. Digital image by *Mario Merino*.

Moments after it all begin about all that is left of the ring of radiation-spreading ring of fire is a few whiffs of smoke here and there, yet no apparent physical destruction and no ground fires seen at ground zero. Digital image by *Mario Merino*.

No physical destruction and no ground fires....but just what is this strange powder-like stuff floating down like flakes of snow around New York City's Empire State Building...all around Manhattan? Digital image by *Mario Merino*.

The pilot of the winged *Sänger* orbital rocket bomber is contemplating his/her landing approach in *Nazi* Germany, unaware of what, if anything, has been accomplished by the radiation-spreading bomb released 78 miles above Indianapolis, Indiana, and 3,500 miles away. Digital image by *Mario Merino*.

The winged orbital bomber is now gliding silently home at about 25 miles [40 kilometers] altitude preparing to land at its *Luftwaffe* air base. A solid cloud mass covers the Earth as it prepares to drop below its current 25 mile altitude and somewhere below is the shore line of the Atlantic Ocean. Digital image by *Mario Merino*.

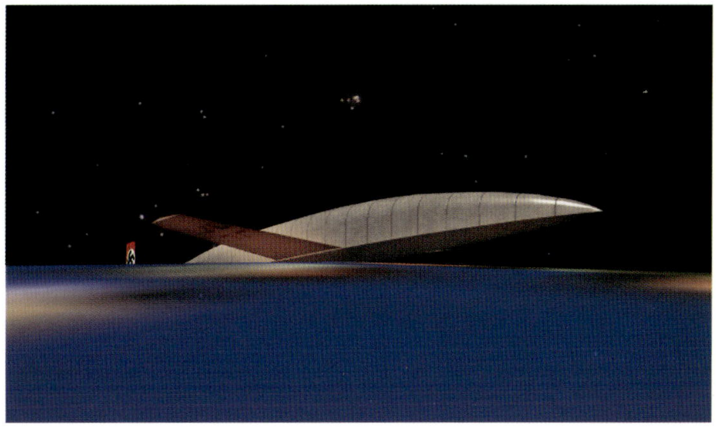

Suddenly the white cloud cover is gone. Digital image by *Mario Merino*.

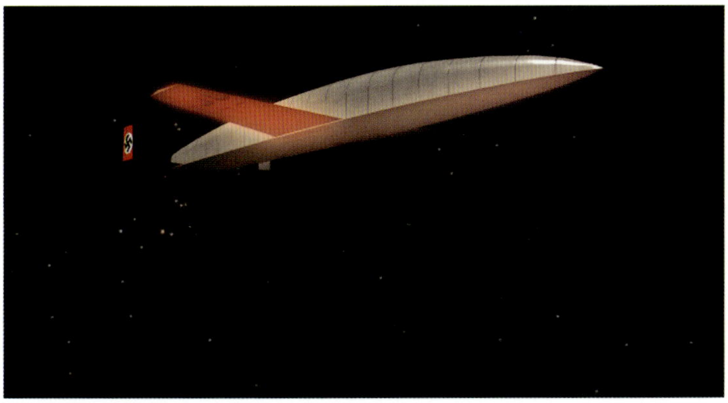

As the *Sänger* winged orbital bomber flies through the dense atmosphere friction increases, and the flying machine's surfaces are heating up again. Digital image by *Mario Merino*.

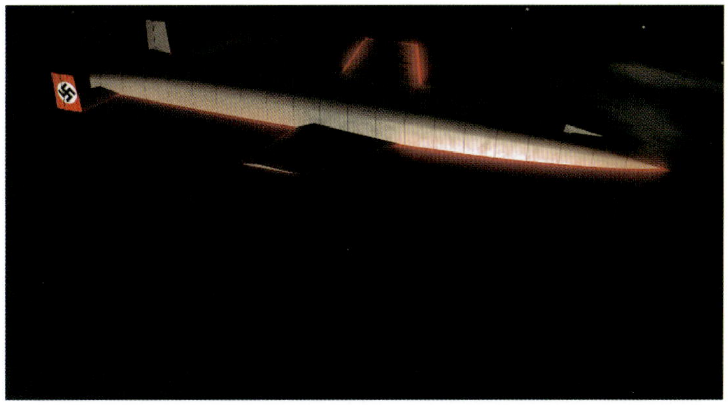

Empty. The *Sänger* winged bomber doesn't weigh much now...between 10- and 11-ton (20,000 to 22,000 pounds). Digital image by *Mario Merino*.

A *Sänger* orbital bomber, rapidly descending, is beginning to glow all over, passing through the dense upper atmosphere on its way down to a landing site in northern Germany. Digital image by *Mario Merino*.

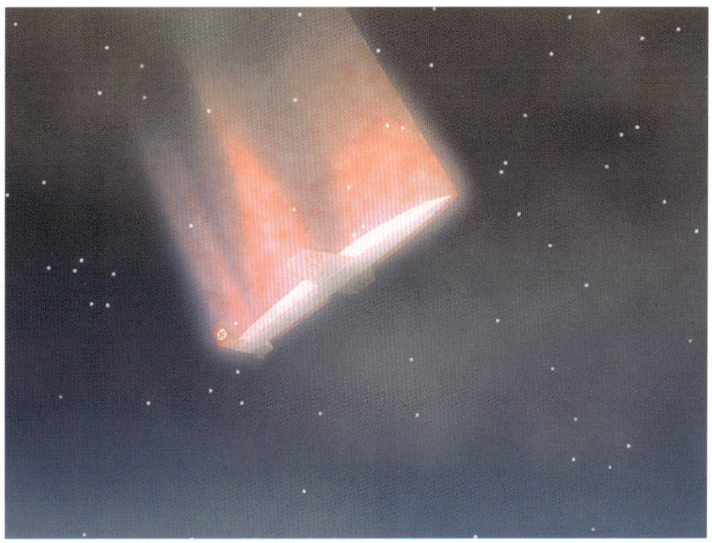

The *Sänger* orbital bomber, its mission completed, continues to pass through the dense atmosphere on its way back to earth while its surface is grows hotter and hotter. Digital image by *Mario Merino*.

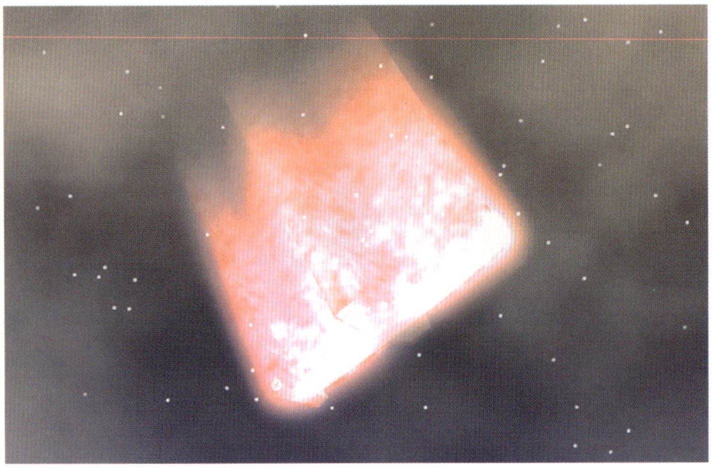

It appears that the friction of the dense atmosphere is going to be to much for the reentering *Sänger* orbital bomber. But it will survive being engulfed in flames. Digital image by *Mario Merino*.

The *Sänger* orbital bomber is now in view from the ground having come through the burning friction of the dense atm0spshere. Digital image by *Mario Merino*.

With its landing flaps down, the *Sänger* orbital bomber is anticipating its landing back on German soil. Digital image by *Mario Merino*.

Landing gear doors on the underside of the *Sänger* orbital bomber are seen opening in anticipation of unfolding its tricycle undercarriage. Digital image by *Mario Merino*.

Full flaps, a nose wheel, and two single wheeled main wheels are extending out of the *Sänger* orbital bombers fuselage. Digital image by *Mario Merino*.

The landing gear on this *Sänger* orbital bomber is seen full out and locked. Digital image by *Mario Merino*.

A nose on view of a *Sänger* orbital bomber manuvering down toward its landing strip. Digital image by *Mario Merino*.

A nose port side view of a *Sänger* orbital bomber seen high over the German country side with its landing gear full out. Digital image by *Mario Merino*.

The angle of attack of this *Sänger* orbital bomber is steep meaning that it is seeking to slow its forward motion by creating drag. Digital image by *Mario Merino*.

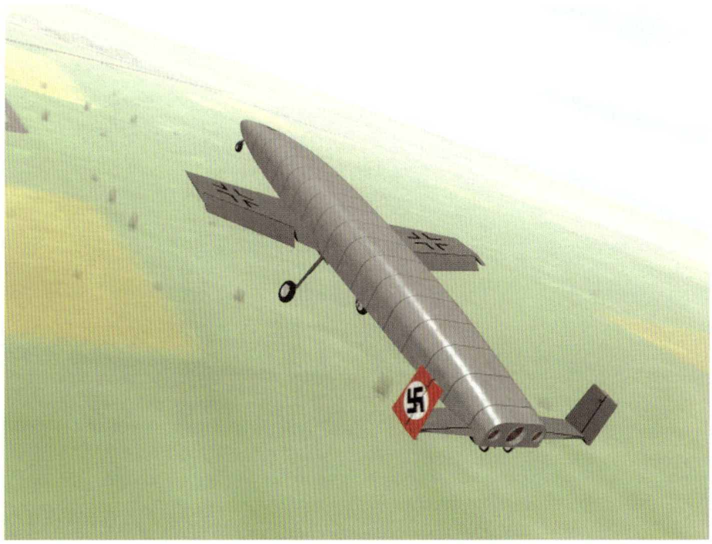

This *Sänger* orbital bomber is a glider at this point...having to thrust to maneuver in anticipation of landing. Digital image by *Mario Merino*.

Finally, the pilot of this *Sänger* orbital bomber see his/her landing strip off in the horizon. Digital image by *Mario Merino*.

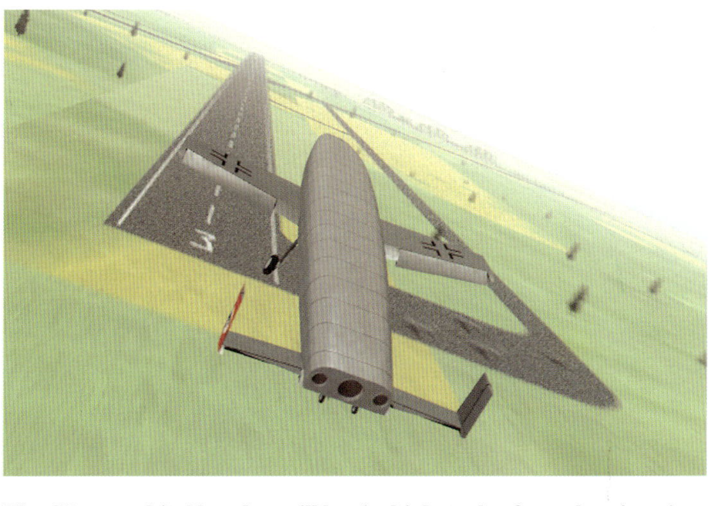

The *Sänger* orbital bomber still has its high angle of attack as it strives to line up with its landing strip. Digital image by *Mario Merino*.

This *Sänger* orbital bomber appears perfectly lined up over the white center line of its landing strip somewhere in northern Germany. It is almost home. Digital image by *Mario Merino*.

Eugen Sänger designed his orbital bomber to be a tail-dragger...believing that it would help slow the flying machine as it completed its landing run. A nose wheel was built in to keep the orbital bomber from nosing down upon landing. Digital image by *Mario Merino*.

Touch down for this *Sänger* orbital rocket bomber as tire smoke and water vaporize behind the 92 foot long 10-ton empty weight piloted bomber. Digital image by *Josha Hildwine*.

A nose port side view of an idle *Sänger* orbital rocket bomber supported by its long legged tricycle landing gear. Another day and another *Sänger* orbital rocket bomber is ready for another flight across the Atlantic Ocean. Digital image by *Mario Merino*.

Appendix A

Mission Profiles of the *Sänger* Rocket Bomber and the Wide Variety of Global Bombing Attacks its Global Reach/Range Now Made Possible

01. Two Basic Types Of Bombing Attacks:
Single Point and Wide Area

The individual types of a "point" bombing attack follow from the requirement that the rocket bomber fly relatively slowly over the target to enable accurate bomb strikes. In such cases, the vehicle so that it may have rather small residual energy. there. If, in spite of this, the bomber is then to return to its takeoff field after dropping its bombs over the target and without a refueling stopover, then it must be propelled by its own motor (a second rocket 'burn') until it has acquired sufficient speed and energy to get home. on the corresponding energy. Thus he *Sänger* arrived at theorized procedures for a bomb attack involving two rocket burn sequence times and a return home. This method consisted of having , which consists essentially in having the bomber, after being rocket catapult-launched at the home airbase, accelerate only until it acquires enough speed and energy to bring it over the intended bombing target. Then the rocket bomber would release its bomb weapon payload and turn, at the lowest possible speed, back toward its home airfield. Then the pilot would restart the its rocket motor with using the residual remaining stores of rocket fuel, get up accumulate sufficient speed and energy for the cruise home, trip, and finally land back at his the home airbase. Very large quantities of fuel would have been required for this double rocket burn, time, so that the entire this type of bombing operation could would have been used only for limited ranges of bomb attack up to 6,000 kilometers [3,728 miles]. The and a limited tonnage of the bomb load on such "dual burn" flights also would have been limited. Point bombing attacks for over greater distances, or with target bomb loads greater heavier than the example just mentioned, would be possible, but only if the rocket bomber could land at an alternative "friendly" site not too far from the bomb principal intended target and take on a fresh load of fuel.

For the bombing attack involving two separate rocket burn times, partial turning, and landing at an auxiliary airfield, the *Sänger* rocket bomber would again be accelerated for the 2nd time after its initial catapult-launch from its home airbase, until it had acquired sufficient forward motion to carry it to the intended target. There it would release its bomb load, begin turning through the required angle offering the least possible loss of flight speed, start its rocket motor with the small residue of remaining fuel on board, to get another shot of energy which would carry it on to and reset a course to an auxiliary airfield not too far distant from the planned target area. which it had just bombed. The rocket bomber would then land and take on new load of fuel. Once refueled, the rocket bomber would immediately take off again in its normal catapult-launch fashion returning and return to its home airbase in Germany. In addition, were If the rocket bomber to could also take on a fresh load of bombs, too, it would then have the possibility of making further bombing attacks on its way home.

If a "point" bombing attack was to be carried out over a long distance, or with a very great bomb load, and there was no possible auxiliary landing airfield near the target, then there might be no possibility of retrieving the bomber – that is, bringing it back to its home airfield. If the attack on the target seemed more important than the bomber aircraft itself (which had only relatively small material value, said *Sänger*), then there was the possibility the option existed of sacrificing the rocket bomber after the mission was carried out. sometime after completing its bombing raid. This procedure of conducting a point bombing attack with a single rocket burn time and then sacrificing of the rocket bomber was, in principle, applicable to all point bombing targets anywhere on the Earth's surface. They This option could naturally to be applied to attacks on and targets of special significance as, for example, like the surprise destruction of a specific government building and the governing group assembled there (like the U.S. Capitol); the assassination of a single, especially important enemy leader (if his whereabouts was known); the sinking of large enemy transports or ships of war (like an aircraft carrier task group); the blocking disabling of important avenues of commerce (like the locks of the Panama Canal); canals or straits) and to similar special cases. Such missions would not be considered 'suicidal', for there would be a means of ejecting the pilot's cockpit capsule from the vehicle and returning him to the ground by parachute. Assuming the pilot would land uninjured, he might be able to find his way back to German territory. ; this is less because of the loss of the aircraft than the more valuable pilot.

For procedures of an "area" bombing attack on a wider region, area, such as a city of one million population, the need to fly slowly over the target to enable highly accurate bomb placement – as in the point bombing example – disappears. This would give so that the pilot has more freedom options and flexibility in the release of the bomber's ordnance. carrying out the procedure. The most obvious procedure for the an area bombing attack, with single rocket burn time and the immediate return to home airbase, consisted of the rocket bomber being catapult-launched from its home airbase and then powered only long enough until it got to gain sufficient speed and energy to get to the vicinity of the bomb target. Then the aircraft would turn around and be flown back to its home airbase. Complete turn around would use up very large amounts of rocket fuel so that this bombing attack would be limited to relatively short flying distances and small lighter bomb loads.

As in "point" bombing, an "area" bombing attack over great distances would be very simple easier if an auxiliary airfield existed not too far from the primary target. This way, the rocket bomber could land and take on new fuel and perhaps bombs, too, for the return trip home. If some targets happened to be lie on the rocket bomber's return flight path, these could be struck as well. well, it could drop them. Nevertheless, the course of events in an "area" bombing attack would have played out as follows:

• After bomb release, the rocket bomber would make a partial turn through an angle less than 180 degrees (this required less rocket fuel than would a complete 180 degree turn) and then on to accelerate – using a second rocket burn and its remaining rocket fuel – toward its an auxiliary airfield.

• An area bombing attack carried out at longer range, using with a single rocket burn, time, a partial turn and an auxiliary airfield landing, would be applicable to virtually all potential target sites/distances on earth.

• It matters, however, In many potential target scenarios, there would be a suitable friendly civilian or military airfield for landing, refueling, rearming and relaunching the rocket bomber within refueling equipment, and a catapult-launch takeoff monorail apparatus the existence that within at most a few thousand kilometers [1,000 to 1,200 miles] from the of the principal bomb target. that;

Although it appeared that a large number of worthwhile targets existed throughout the world for potential offensive area bombing attacks using the rocket bomber, *Sänger* wrote that area bombing might only be used in exceptional cases. He did not elaborate.

The value of an auxiliary alternative target could vary considerably, not only according to its distance from the principal target, but also because of the size of the required angle of turn needed for it's the bomber's return flight home. Since large angles of turn require much more fuel than even long distance flights, an obvious idea of the rocket bomber's planners was to provide auxiliary airfields near all foreseeable targets, that is, beyond the two population concentrations outside of Europe (North America and South East Asia) such as in the Marianas in the Pacific Ocean, the islands off the Mexican coast, or a secure auxiliary airfield at the antipodes of the home airbase, say in New Zealand, or on the islands east of New Zealand. This auxiliary airfield at the antipodes could always be reached by straight flight without turning, no matter what point on earth the rocket bomber had just attacked. The rocket bomber's distance from the target could be very distant from the bombing. *Sänger* wrote that his thinking was based on the method of attack with a single rocket burn time and a auxiliary airfield at the antipodes. Such a single auxiliary airfield had the advantage, he said, that it could easily be fully equipped with fuel and bombs to enable rocket bomber to make additional bombing raids on the journey back to its home airbase. If the auxiliary airfield's island location could be easily protected against enemy attacks by air or surface fleets, perhaps a 2nd rocket bomber might attack the enemy airfields and ship yards before the rocket bomber was due to land.

If such an auxiliary airfield in the antipodes would not be available because it could not be easily protected against enemy attacks by air or surface fleets, then area bombing attacks over long distances could be carried out by having the rocket bomber, after the release of its bomb load over the primary target, fly a straight course (circumnavigation) all the way around the Earth until it reached its home airbase. This procedure would be known as an area bombing attack with single rocket burn time and then circumnavigation of the globe.

Sänger summarized that all possible procedures for an area bomb attack were all the same for the point-attack regarding what happen up to the moment of bomb release. However, post bombing procedures differed only in the manner of bringing the rocket bomber back home after it released its initial bomb release.

02. Point Bomb Attack - Two Separate Rocket Burn Times, Full Turn Around, and Return To Home Airbase

Sänger stated that a point bombing attack up to a radius of 2,000 kilometers [1,243 miles] could be achieved with his rocket bomber. This would include valuable targets such as:

• All targets throughout England and Europe;
• All targets between Moscow, USSR and Madrid, Spain;
• All targets between northern Sweden and Tripoli, Liberia;
• All targets between Ireland and the North Pole;

Sänger wrote that any of these bomb targets could be hit with extreme accuracy, that is, whether it be any object on land or sea, with a bomb load as much as 30-ton and no matter how small in size the target might be. He assured his readers of his proposal that 30-tons of bombs would destroy all but especially the most hardened concrete structures within 300 meters [984 feet], penetrate earthen works 100 meters [328 feet] thick, and steel armor plate one meter [3 feet] thick. The *Sänger* rocket bomber would then return to its home airbase without a stopover at some auxiliary airfield. On the other hand, *Sänger* stated that with a bomb load of 15-tons, the same rocket bomber could carry out attacks to over roughly 6,000 kilometers [3,700 miles] from its home airbase. At this range, the rocket bomber could bomb strike targets between:

- Germany and Central Africa;
- Germany and Hindustan;
- Germany and Eastern Siberia;
- Germany and all the North Polar Region;
- Germany to the Atlantic coast of North America;
- Germany and the whole of the North Atlantic Ocean;

The procedure of a point bombing attack with two separate rocket burn times appeared to *Sänger* to have extraordinary practical importance. and He felt that it would be applicable to all the *Luftwaffe's* offensive operating areas actions inside Europe...even neighboring regions. Even though during its point bomb attack, the rocket bomber might come into an enemy's defensive zone due to its relatively low altitude and low speed, it would have the value of complete surprise in its favor, said *Sänger.* Because of the projected altitude and speed of the bomber's operation, it would be virtually invulnerable to all defensive systems known and operational at that time. was the fact that the rocket machine would be a complete surprise that even a warship on full alert would not have sufficient time to harass the rocket bomber, warding off its attack and driving it away, or much less in shooting it down.

03. Point Attack - Two Rocket Burn Times, Partial Turn, and Landing at an Auxiliary Airbase

This method of bombing attack differed from the one mentioned only because the return flight of the rocket bomber could be shorter than its outward trip from the home airbase to the target. A landing is to be would be made at a suitable auxiliary location other than home base. *Sänger* said because of the small kinetic energy over the target, the angle of turn was unimportant. The only important factor was the flying distance from the bomb primary target to the auxiliary airfield.

An example of a point bombing attack with two separate rocket burn times and an auxiliary airfield (near the target) would have been an bomb attack of the locks of the Panama Canal with a landing at an auxiliary airfield on the west coast of South America. *Sänger* believed that this type of attack would only be sensible for those parts of the Earth's surface primary targets which were nearer to the auxiliary airfield than to the home airbase of the bomber. This possible range of attack included all of East Asia and a large part of the Western Pacific. The antipodal auxiliary airfield also would be important for alternative area bomb attacks, its value, only in point bombing attacks, was when the if the defense of a specific primary "point" target itself was protected increased after launch of the bomber (say against attacks on a navy fleet). by rocket bombers from the home airbase. Specifications for this type of bombing attack included:

- Bomb load - 2-ton;
- Bomber takeoff weight - 100-ton;
- Climb from catapult-launch takeoff monorail to a northeasterly direction - 11 seconds;
- Weight of bomber after liftoff from rocket sled - 100-ton;

End of motorless flight:
- Total elapsed time - 36 seconds;
- Velocity - 284 meters per second;
- Altitude - 3.7 kilometers;
- Distance traveled - 12 kilometers;

End of powered climb period:
- Time elapsed - 332 seconds;
- Weight of bomber (with payload) - 36 ton;
- Velocity - 4,560 meters per second;
- Altitude - 60 kilometers;
- Distance traveled - 512 kilometers;

End of supersonic descent:
- Total elapsed time - 3,882 seconds;
- Weight - 8-ton;
- Velocity - 300 meters per second;
- Altitude - 14 kilometers;
- Distance traveled - 9,390 kilometers;

End of the subsonic descent ends with the start of the diving attack; the final altitude would be determined by the succeeding dive. *Sänger* wrote, for example, that for an attack on the Panama Canal locks, that maximum accuracy of hits would be more important than high impact velocity of the bomb. The end of the subsonic descent chosen would be 2 kilometers [1 1/4 miles] altitude. Other numbers include:

- Time elapsed - 4,162 seconds;
- Aircraft weight - 26 tons;

- Velocity - 142 meters per second;
- Altitude - not available;
- Distance traveled - 9,450 kilometers;

End of dive attack. The dive attack would have gone from an altitude of 2 kilometers to about 0.5 kilometers;

- Time elapsed - 4,405 seconds;
- Weight - 24-ton;
- Velocity - 200 meters per second;
- Altitude - 0 kilometers;
- Distance traveled - 9,450 kilometers;

Once the rocket bomber's bombs had been released, the machine would have suffered a small loss in velocity, approaching to within negligible distance from the earth's surface.

- Time elapsed - 4,172 seconds [7 minutes];
- Weight - 24-ton;
- Velocity - 200 meters per seconds [447 miles/hour];
- Altitude - 0 kilometers;
- Distance traveled - 9,450 kilometers [5,872 miles];

End of 2nd Climb Period:

- Time elapsed - 4,405 seconds [73 minutes];
- Weight of aircraft after second rocket burn - 10-ton;
- Velocity 2,800 meters/second [9,186 feet/second or 6,263 miles/hour];
- Altitude - 22 kilometers [13 1/2 miles];
- Distance traveled - 9,710 kilometers [6,034 miles];

End of 2nd Subsonic Glide-Flight:

- Time elapsed - 6,785 seconds [113 minutes];
- Weight pf aircraft - 10-ton;
- Velocity - 80 meters/second [262 feet/second or 178 miles/hour];
- Altitude - 0 kilometers;
- Distance traveled - 12,648 kilometers [7,860 miles];

Landing:

- Time elapsed - 6,810 seconds [113 1/2 minutes];
- Weight off aircraft - 10-ton;
- Velocity - 0 meters per second;
- Altitude - 0 kilometers;
- Distance traveled - 12,650 kilometers [7,457 miles];

04. Point Bomb Attack - Ditching the Rocket Bomber after Releasing its Bomb Load

Sänger claimed that his rocket bomber with a moderate exhaust speed could carry out attacks against small individual targets up to 6,000 kilometers [3,700 miles] distance from its home airbase. If there was an antipodal auxiliary airfield, its range could extend up to 2,000 kilometers [1,200 miles] more beyond the first primary target. If however, the "point" bombing attack was to be directed at a primary target in whose vicinity there was with no auxiliary landing place, in the near vicinity, however, there was the possibility that the bomber could proceed exactly as planned and land at a corresponding airfield near the target even though the technical installations of an auxiliary airfield did not exist there. In this case, it the bomber would no longer be able to take off from this auxiliary site under its own power, but would not necessarily would be lost if the a landing did not occur in friendly, but rather occurred in enemy territory. If there was no other possibility than landing in enemy territory, then we would be left, said *Sänger* with a last conclusion, not to be neglected — This would mean a deliberate point bombing attack with single rocket burn time and sacrifice of the rocket bomber and probably its pilot.

Since the rocket powered aircraft would gain altitude rapidly after a bomb release in a point bomb attack, the pilot could – at the end of this brief climb – parachute from the rocket bomber and allow the machine to crash, down destroying itself in order to keep it from following rather than falling into the hands of the enemy. The pilot would likely land several kilometers [miles] from the site of the bomb impact and in all likelihood, be captured. *Sänger* said that from a point of view of performance this procedure represented a limited case where of point bomb attack where the return fight would be zero but the attacking range covered the entire surface of the Earth. Delivery of a bomb load , say of a half ton [1,000 pounds] could be carried out at to points most of maximum distance from its home airbase. Were the rocket bomber to enjoy a velocity of 4,000 meters/second [2 1/2 miles/second], its bomb load could be increased to 8-tons and with at a forward velocity of 5,000 meters/second [just over 3 miles/second], the weapon payload could rise to 17-ton. *Sänger* also mentioned that this procedure would be naturally suited to an unmanned use of the rocket bomber, but he didn't elaborate.

05. Area Bomb Attack - Full Turn Back to Home Air Base After Release of Bomb Load

This area bombing attack is different from the other examples in that because the quantity of rocket fuel loaded on board at the beginning of the flight would have to suffice for the outward flight and the entire return trip back to the home airbase. No stopovers. A large substantial amount of kinetic energy would be present over the bombing target, but considerable percentages a large portion of this energy would be lost in turning making a turn for the return trip home. It had to be assumed that the supersonic descending flight path, both during the segment flight before bomb release and then during the bomber's turn, would be carried out at a stationary altitude rather than in the a strongly oscillating dynamic flight path. This would have been necessary for the decisive third phase of aiming in order to release the bombs during horizontal flight and attain the necessary aiming accuracy for the bomb release hitting the target. and in order. Stationary Constant altitudes of flight would also would be necessary for the period the duration of the turn in order to set up the predict the required aerodynamic turning forces.

This last task – turning the machine around – involved required from the rocket bomber's pilot would be the fact that turning the flying machine around would be possible only up to at certain forward speeds. If the rocket bomber has a high forward speed was excessive, then other methods of bombing attacks would have to be used because the great its forward speed the less it would be able then the bomber would be unable to make a 180 degree turn without aerodynamically destroying itself in the process. *Sänger* believed that area bombing attacks, with a single rocket burn time, and full turn would be limited to flight ranges up to 12,000 kilometers [7,500 miles]. Inside the this 12,000 kilometer range, area bombing attacks would be offer a very effective use of his rocket bombing machine, despite the very costly somewhat difficult turning process required in order to return to it's the home airbase. Beyond 12,000 kilometers, [7,500 miles] the rocket bomber might not make probably would not have sufficient fuel to get it back home if making a full turn. plus a full turn..

As an example of an area bombing attack with single rocket burn time and full turn, *Sänger* gave as an the example of a rocket bomber attack on New York City. By launching westward, this target would be 6,500 kilometers [4,000 miles] from the rocket bomber's home airbase. This would be well within its range, including a full turn. a round. So if the rocket bomber had a forward velocity of 4,000 meters/second, or almost 2 1/2 miles/second . In addition, it could carry a bomb load of six tons and still make a full turn and return home after bombing New York City. The following are *Sänger's* details on how a bombing attack on New York City would have been carried out by his rocket bomber in an area bombing attack:

• The rocket bomber's rocket motor is fired up at 36 seconds after takeoff and approximately 12 kilometers [7 1/2 miles] from its takeoff catapult-launch ramp;

• Within the next 336 seconds the rocket bomber would consume its total on board fuel supply of 84-ton;

• At the end of the rocket bomber's climb, it would have reached a forward velocity of 6,370 meters/second or almost 5 miles/second hour;

• At an altitude of 91 kilometers [56 1/2 miles], it would have traveled a total of 736 kilometers [457 miles] from its takeoff point in Germany;

• At this time the rocket bomber would now weigh approximately 16-ton;

• Using only its store of potential and kinetic energy, the rocket bomber would continue to fly on to New York City, its target, approximately 5,550 kilometers distant, having flown 950 kilo- meters and an elapsed time of 11,050 seconds from its home airbase;

• The rocket bomber's velocity would have dropped to meters/second and a stationary altitude of 50 kilometers [31 miles];

• The bomb(s) is/are released;

• After bomb release, the rocket bomber weight would be 10-tons;

The rocket bomber, after its bomb release on New York City, would go into enter a full 180- degree turn and in approximately 330 seconds, goes through a turn spiral of roughly 1,000 kilometers [621 miles] in diameter, until it had reached would reach the right compass heading/direction for its return flight to Germany. During this turn, turning, its altitude would be decreasing in order to develop the preserve the kinetic energy and remain inside acceptable aerodynamic forces necessary for the turn. At the end of the rocket bomber's turning maneuvers, its forward velocity would have declined to approximately 3,700 meters/second and the machine would have dropped to an altitude of 38 kilometers [23 miles]. The rocket bomber's supersonic glide-path now would place it puts it in the direction of its target home airfield, which is lies approximately 5,459 kilometers [3,400 miles] away. It would take it approximately 2,600 seconds [43 minutes] to get there. Its glide-path would end 100 kilometers [62 miles] before landing at its home airbase. At this time the rocket bomber would be flying without power at approximately 20 kilometers [12 1/4 miles] in altitude and a forward velocity of 300 meters/second [984 feet/second]. [Reader, it may be a bit hard to 'take in' as the L/D at the end of the turn would have to be about 1:200 to get the vehicle back to German airspace – 62 miles from TD at 60,000 feet for entry to 'final glide.'. Even considering the near vacuum at 100,000 feet altitude at the 'entry' of the return glide – and assuming no fuel remains on board for a second burn – are these kinds of numbers rational here for an unpowered glide over a 3,400 mile distance. That's the equivalent of flying across the United States, beginning at 100,000 feet with no power, assuming an entry speed of 3,700 meters/second...maybe 2.5 miles/second.

06. Area Bomb Attack - Partial Turn and Landing at an Auxiliary Airfield

This type of bombing attack corresponds to the point bombing attack but with two separate rocket burn times, partial turn, and landing at an auxiliary site. The same procedure could be followed, however, with only one rocket burn time.

This bombing procedure could have been used to bomb the west coast of America, *Sänger* said. This was true because in this case, the rocket bomber's turn angle was considered to be virtually zero. There was a definite sized bomb load of 5-ton with which the rocket bomber could deliver on the way to its auxiliary landing airfield. So it the bomber could drop release its bomb load at any point along on the way to the auxiliary landing site without altering its original range. This, of course, assumed provided the bomb release occurred after the bomber's rocket engine was off so that the range breadth of area for this area bombing attack was somewhere between its takeoff point in Germany and an auxiliary landing site. With bomb loads under five-tons, the possible area of attack included all of western North America, plus considerable portions of South America and the Pacific Ocean.

07. Area Bomb Attack - Antipodal Auxiliary Airfield Landing

When the area to be bombed is antipodal to the rocket bomber's home airbase, then there is no need to turn the rocket bomber around and no loss of energy turning the machine around at speeds above the speed of sound. The rocket bomber's flight would be rectilinear with a total distance to the landing site of somewhere around 20,000 kilometers [12,400 miles]. Areas to be bombed would be arbitrary and independent of its proposed 0.7-ton bomb load. and It also wouldn't matter at what place on along the rocket bomber's glide-flight path its bombs would be released.

08. Area Bomb Attack - Full Global Circumnavigation... Home Airbase to Home Airbase

Sänger said that bombing with a full global circumnavigation bombing most clearly demonstrated the extreme technical superiority of his rocket bomber. showed, which With a size and empty weight equal to that of a contemporary medium military bomber, aircraft his rocket bomber could reach every point on the Earth's surface with a bomb load of 3-tons and with at moderate exhaust speeds. It would be able to fly up to 40,000 kilometers [24,800 miles], fully the complete distance around the world without an intermediate landing.

One example of an area bombing attack with a single rocket burn time and circumnavigation would be bombing attacks on cities of one million or more people most distant from Germany.

Appendix B

The Evolution of the Lifting Reentry Concept, Beginning with *Eugen Sänger*

Theoretical Foundations

Tsiolkovskiy, *Oberth*, and *Goddard* all undertake generalized studies of rocket-propelled spacecraft.

Reentry Requirement Studies

Valier projects concept of winged reentry. *Hohmann* studies orbital transfer and reentry mechanics.

Early Reentry Vehicle Conceptions

Sänger and *Bredt* advocate a "Rocket Space plane" (1930-1944).

Development of the *Peenemünde A-4b* and demonstration of supersonic glide (1945).

Early Post WWII - Interest

Sänger-Bredt report inspires studies of lifting reentry by *Rand*, *Caltech*, *Bell*, *Douglas*, SAB, *ONR*, and *NACA*.

The 1950s: Formative Years

Woods-Dornberger-Stack influence inspires *NACA-USAF-Navy X-15* program.

L. R.. Carman and *H. Drake* de-*TSTO* reentry vehicle.

von Braun, drawing on *Peenemünde* work, postulates a series of winged rocketry vehicles for orbital and interplanetary missions.

Winged reentry research begun at *NACA*-Langley.

H. Julian Allen postulates blunt body concept, leading to the first practical reentry vehicles.

Bell, undertakes *Bomi* study, inspiring *118P, Brass Bell, Robo,* and *Hywards* undertaken subsequently by *Bell* and other contractors for the *USAF*.

Initial Reentry Vehicles

Allen research leads to *Atlas* warhead reentry vehicle, and influences *Faget* design of *Mercury* ballistic spacecraft.

Allen research spans the *M* series of lifting bodies, suggest for *LUNEX*.

Love research leads to *HL-10*

USAF consolidates *Robo*, *Brass Bell*, and *Hywards* into *Project Dyna Soar*, leading to *Boeing X-20A*, canceled in 1963.

USAF initiates *START*, consisting of *ASSET*, *PRIME*, and *PILOT*, demonstrating maneuvering lifting reentry and leading to the *Martin X-24A*.

NASA undertakes development of low-speed *M2-F1, M2-F2*, *M2-F3,* and *HL-10* testbeds.

1960s Pre-Shuttle Studies

USAF "*Aerospaceplane* project.

NASA studies for orbiting space station supporting by logistical lifting reentry spacecraft capable of fulfilling space rescue, by *Aerojet*, *Boeing*, *Douglas*, *General Dynamics*, *General Electric*, *Lockheed*, *McDonnell*, *Martin*, *North American*, and *Northrop*.

European studies by *ERNO*, *ONERA Sud*, *Nord*, *Dassault*, *SNECMA*, *Hawker-Siddeley*, *Junkers*, *RAE*, *Messerschmitt-Bölkow-Blohlm*, *Dornier*, and *Aerospatiale.*

Joint *NASA-DoD AACB* report (1966).

Shuttle Conceptualization

Formation of *NASA Space Shuttle Task Force*.

Award of *ILRV* study contracts to *Lockheed*, *General Dynamics*, *McDonnell-Douglas*, and *North American-Rockwell*. (1969)

NASA-SI International Shuttle Symposium (1969).

Phase B Study Contract award award to *McDonnell-Douglas*, *North American-Rockwell*, plus *Phase A* award to Grumman-*Boeing* and Lockheed (1970).

Apollo CM demonstrates moderate *L/D* maneuvering reentry.

Development of Space Shuttle Orbiter

External factors influence *Shuttle* design *NASA* cancels space station; *Office of Management & Budget* limitations; *Department Of Defense* needs.

Phase *B* evolves into *B* and *B* studies.

Presidential endorsement of *Shuttle* on January 5th, 1972.

NASA selects parallel-burn partially expendable booster design.

Rockwell derives *Shuttle* orbiter in conjunction with *NASA* and *JSC*.

Rockwell selected over *Grumman*, *McDonnell-Douglas* and *Lockheed* by a joint *NASA-USAF* Source Selection Board (July 1972).

Construction of *Shuttle* begins (1974).

First approach and landing tests (1977).

First orbital flight (1980).

Index